The New DNA

Personal and Universal Transformation

by Jayne Chilkes

Avid Readers Publishing Group
Lakewood, California

The New DNA Personal and Universal Transformation

Avid Readers Publishing Group

http://www.avidreaderspg.com

ISBN-13: 978-1-935105-50-3

Printed in the United States

Works by Jayne Chilkes

Books
The Call of an Angel
Twelve Steps to Heaven

Published Articles
Over 21
Elite Chic
Soho News
ID Magazine
Time Out
Planet Lightworker

Music CDs
Land of Poetry
Inner Knowledge
The Call of an Angel
The Dance of Life
The Call of an Angel Visualisations I and II

http://theangelschannel.netfirms.com

Jayne Chilkes (UK) ALCM, MABCH, ITEC, LOPC is an experienced Healer, Soul Channel and Expert in Twin Flame Reunion, Hypnotherapist, DNA Practitioner, Coach, Teacher, Author, Soul Artist and Composer. She has worked in the Healing Arts for nearly thirty years.

She currently lives in the USA and offers Soul Channelling, DNA Activations, Akashic Record Clearing and Workshops internationally.

Her main goal is to share Peace, Love, Wisdom, and New Earth information that she has received, with All.

Thank you to my dear husband Richard Quataert, Maurene Watson, my Twin Flame, my Soul Group, and all friends and family who have inspired me to write this book.

Preface

Greetings,
The world and our lives are in the epicentre of massive change. We are in the midst of all old systems disintegrating before our eyes, yet many remain unaware because the matrix has imprisoned their consciousness.

At the same time, there is the potential for a great spiritual awakening, which can transform the earth into a harmonious existence amongst all living beings. For us as humans, this shift is occurring within our DNA accelerating us towards the next stage of our evolution.

I received a DNA Activation in 2004, which had a dramatic and powerful impact on my life. This information has been an important message of positive change and awakening, which I have been blessed to receive and share with others.

DNA Activations and Akashic Record Clearing has propelled many forward and made people more ready as we rapidly approach and live in the vibrations of the New Earth.

This book will include suggestions of how to open up and live in a new paradigm and leave the old matrix behind us.

With much love to you all, as we remember our beautiful world and our amazing human potential.

Abundant blessings
Jayne Chilkes

Contents

The New DNA
Personal and Universal Transformation

by Jayne Chilkes

List of Illustrations

Chapter One

UNDERSTANDING SOUL MATES

There is a whole world of difference between your Soul Mate and Twin Flame. It is said that we could meet about twelve soul mates in one lifetime which can be friendships or personal relationships. The soul mate can be challenging or peaceful depending on your karma with each other, but certainly a very strong bond indeed. So many lessons can be learnt from your soul mate if you are willing to learn them!

Some soul mate relationships can be very confusing. The past lives between each other are intertwined with the present, and whilst the memories from the past intrude into the present, it can be very hard to unravel the puzzle. You may never resolve your issues but it still is a learning curve. Some of these relationships were never meant to be resolved and the biggest lesson is to accept, forgive and let go.

Soul mate relationships can last a few days, weeks, months or a lifetime. It is not the length of time that is important but the quality and learning that you both need to receive. Also your soul mate is not necessarily a perfect person (just like you are *n*ot a perfect person) to fulfill every expectation, or delivering every desire once you are both together. You are participating in a wonderful growth process and opportunity for more spiritual awakening.

Communication and forgiveness is the key and solution for moving forward whatever the situation and of course unconditional love to boot. If things get really difficult, I recommend the two main work sheets from the book <u>Radical Forgiveness</u> by Colin Tipping at www.radicalforgiveness.com.

Here is my story of meeting my soul-mate/husband/friend,

especially written to inspire people that things can work out and all things are possible. This book begins where I left off from my last book "Twelve Steps to Heaven."

Never Say Never – My Journey

"I was brought up in the UK just outside of London in a happy home with creative parents. I learnt to play the piano at the age of six and went to art school in my teens, becoming a fashion designer and a part of the New Romantic fashion scene in the 1980's with Boy George, Spandau Ballet and the like. Since then, I have worked in the healing arts for over twenty years and I have to say my relationship pattern during the majority of my life was pretty awful! I had at least fifteen relationships that were particularly painful in various ways - I was mentally abused or two-timed and shown very little love or caring. I worked on loving myself, healing myself, affirmations, hypnotherapy, past life therapy, receiving healing, counselling, visualisations, meditation, and...guess what? My relationships were still awful! I had numerous realizations about why I drew these patterns to me, but I just couldn't seem to break from them.

At age 42, in despair, I gave up! I said "no more." I would rather be alone than go through any further painful scenarios. I was happy to be on my own. Spiritually I grew and for three years received fantastic weekly soul channellings and training that gave me spiritual sustenance. I basked in the light of soul awareness and I met my Twin Flame who was in soul. I had no idea what was meant by a "Twin Flame," however, I always thought I would meet this perfect mate on the earth plane. Well that was not the plan for me, and my human self rebelled as my teacher, Pam, would bring my beloved Twin through in the channellings. Who was he? Why wasn't he on the earth? I didn't like someone loving me so much when I didn't know who he was! Then slowly, it dawned on me. I did know Him and I did love Him just as much as He loved me.

I felt my Twin in my vibration and believed I was all set for Life. Until gradually, over a period of years, my human self really wanted to meet a good, kind and sincere man here on earth.

I knew no one could ever be like my Twin, moreover, the kind of Love given by Him is very private. I knew I needed practical help and loving support in daily living. I was also reading and repeating affirmations daily from Catherine Ponders' book "The Prosperity Secrets of the Ages." One of the primary ones I was affirming stated, "I deserve everything, success is my divine heritage." At that point I was thinking predominantly of abundance, money and success.

One of my friends said "all my friends are having great success in meeting partners at www.match.com - why don't you try it?" "Well," I said "I really am not interested in a relationship right now!" So, very confused, I stepped forward just a little, pretending to myself that nothing would come of it!

I went to the website and I tried the whole of the UK with no one fitting the spiritual understanding I was looking for. Then I saw a great beautiful hunk advertised on the first page of this website from the USA. Does that mean I could get to talk to men in the USA? A light went on as I changed my search criteria and typed in 'over 5,000 miles', and lo and behold up came a list of much more interesting men abroad. There seemed to be men who would actually write paragraphs about themselves and be totally honest. I then found Soulful9. A name that encaptured my 'soul' work and my number! Soulful9 was intelligent, sincere, kind and had been involved with spiritual growth for twenty years.

It was so exciting writing to Soulful9 for the first time! I felt very safe too because he was in Rochester, NY and I was in London, UK. I knew I wasn't going to get physically involved, which would scare me very much. I kept thinking I would like to meet Soulful9 one day, maybe in a year's time. We really got on very well and he wrote so eloquently and profoundly. What was even more bizarre to me was that he worked as a police officer! That seemed a bit dangerous to me, but fortunately he had left working on the street to work in an administrative assignment. Well, a spiritual policeman - that was even stranger!

I kept showing my mother his emails and she seemed to approve. After weeks of this chatting to and fro, sometimes three times a day, I was hooked. We were becoming best friends. We told each other everything about ourselves on a very deep level. That is something very unique about the Internet. You see somebody from the inside first.

Yet, soulful9, whose name is Rick, really didn't seem to make much effort to want to see me or come over to the UK. I later found out he thought it would be impossible to meet. His search was limited to a 25-mile radius and that was it.

His 40th birthday was coming up, so I thought I would surprise him by sending one of my tapes and a copy of my first book "The Call of an Angel" with a photo of myself. I had seen a picture of him, but he hadn't seen a picture of me! His picture, by the way, was absolutely ghastly! Three chins, massive nostrils, glasses and a balding head. I kept reminding myself that this was beyond looks - this was a special friendship.

Then something unexpected happened. After about two months of corresponding, the Universe seemed to have ideas other than our own. A family dilemma occurred where I was living, and I knew I could not be in that atmosphere and continue with my healing and soul channelling work. On the day of the dilemma, Rick wrote to me and said he had received my book and tape and thanked me. He also said that if there was anything within his power that he could do for me, he would. That night I had a vision of knocking on his front door and staying with him for a couple of weeks so I could have a break and re-think. I told him the next morning there was something he could do for me - have me to visit. He immediately took two weeks off from work.

The next day he rang me for the first time and we spoke for over an hour. It was the actual day of his 40th birthday. His voice sounded wonderful to me and I knew I really loved this man. I could hear the kindness and sincerity in his voice. I was very nervous yet excited at my decision, and so was he. As the three weeks passed until my flight to Rochester, he and I grew even more nervous. What did I have to lose? He gave me the telephone numbers of his parents, relations and friends and that made me feel much safer. He reassured me and I just had a good feeling about it all. I knew I was safe and that my soul guides wouldn't let me do anything too stupid. It was so exciting!

I arrived at Rochester, NY airport on June 12th 2000. Rick was so cute - in fact, much cuter than his photo and I was very relieved! There was a rose on the seat of the car and when I arrived at his house there was a huge bouquet of flowers, the biggest bunch of flowers I had ever received. He had bought brand new furniture for my bedroom and I was treated like a queen. We got on so very well, just as we had hoped.

We decided to get married after the first week! We did not want to be apart. Other friends that I call "human angels," arranged it so that I would have no stress or legal problems getting in and out of the country, which was a miracle in itself, and another sign I was meant to stay.

We were married on August 11th 2000 in Lilydale NY, which turned out to be a beautiful sunny day. Lilydale is a spiritual community located in western NY near the city of Fredonia.

I still feel and communicate with my Twin Flame, and Rick and I are very happily married. I have to thank the creators of the Internet, the Universe, and God for such wonderful blessings. I feel I am supported and loved by Rick and his family and, moreover, I learnt that all things are possible. Above all, I learnt never to say never!" © 2000

The reason I am talking about relationships is that no matter how far up the ascension spiral you are, you can still be tweaked and you can still have things to work out. Participating in a personal relationship is the greatest learning curve of all and can speed your own evolvement immensely. When people come to me and say "I have done all the work I need to do," I am always very suspicious! We are human and there is always going to be something to be transformed, moreover, you can still evolve with your human qualities being imperfect. The soul remains intact and perfect and that is really what matters or so I keep telling myself....ha ha ha!

Loving Yourself

It is an on-going process to develop true self-love, one that takes many lifetimes. As we face and release our blocks, we have to leave no stone unturned and clear all issues, thus keeping self-love flowing. If you are looking for a soul-mate relationship, you really need to learn to love yourself more fully, thus, when you do come together with your partner, you have less baggage and issues to resolve, which is always a good idea! The degree that you love yourself will determine your ability to love the other person, who, in turn, will be reflecting back to you many of your own personality traits and qualities, as well as childhood, mother/father issues. It is said that you often choose traits of your mother or father in your partner, both consciously and mostly sub-consciously!

Ways to increase Self Love

- Repeat the mantra "I Love Myself" at least thirty three times a day
- Do things you love to do and feel you deserve it
- Use the daily affirmation "I deserve everything, success is my divine heritage" which covers all aspects of your life
- Allow others in to love you
- Allow yourself abundance in all areas of your life
- Reward yourself and do not punish self
- Let yourself enjoy romance and affection
- Embrace your friendships
- Laugh a lot
- Follow your own intuition
- Turn negative thought to positive affirmations about yourself
- Live your life purpose to the full and love your work
- Pray, meditate and find peace within, on a daily basis
- Care about your body and diet
- Be creative
- Have confidence
- Be your best friend and accept yourself, therefore you will be more accepting of others
- Create a vision board of your perfect soulmate partner in every detail
- Continue to work on your own issues, patterns and programmes
- Build strong boundaries towards those who drain your

energy, or those coming from less than love towards you

- Don't forget to have fun!

Chapter Two

TWIN FLAME DISCERNMENT

There is only one Twin Flame who is your Divine Counterpart and the other half of you. There is no Love so great as that of your Twin Flame and words only touch the surface in describing the depth of Love between you.

Over the years many people have written to me in pain over their Twin Flame, thinking that they have met him or her in the physical realm/third dimension. Yes, people do meet their beloved in the physical, but it is still rare and can be very painful when there is still karma to work through. It is said if 95% of karma on both sides has been completed, that Twin Flames can live happily together.

There is a great deal of confusion over Twin Flames, so I have compiled some criteria below as a way to differentiate between Twin Flames and Soul Mates. It is advisable to be careful and use discernment, whilst extending a period of time to be absolutely positive that this "person" truly is your twin flame. People get obsessed about the Twin Flame concept, when it really doesn't matter too much! That person is there for your growth, along with their own, and that is what is most important.

How to Tell The Difference between Your Twin Flame and Soul Mate

Physical Realm
How to recognise a Soul Mate
It is Common to Meet Your Soul Mate in the Physical

- You may have a life time relationship with your soul mate

- You know the person from other lives
- Please remember there can be many soul mates
- You know it is just another episode in personal growth
- You feel you have met the person before
- You feel you have lessons together
- You feel you have met other soul mates
- Some lessons are challenging and you may need to resolve something
- Its great at first

How to recognise a Twin Flame
You Rarely meet Your Twin Flame in the Physical

- Please remember there is only one true Divine Counterpart offering The Holy Reunion or The Holy Marriage in the physical realm or in soul
- If karma for the female and male is 95% completed on both sides, they have more of a chance to stay together in the physical in this life
- You know the person from the beginning of time
- You know you are in completion
- Feel you know each other inside out and the discovery can go on forever
- Feel complete, yet have some lessons together
- Never felt as deeply before from a soul mate
- Beyond the challenges together you know the connection is unfathomable
- You know your soul is connected forever whatever happens in personality
- There is Male and Female balance in re-union
- You can remember the pain of the split at the beginning of time and upon meeting again heal the memory
- You are supposed to be together to share with the world the perfect union
- You are destined to be spiritual teachers together to teach the world about twin flame union and more

The Subtle Grey Areas
For both your Soul Mates and Twin Flame

- One partner is more evolved than the other
- It is not easy at times, nor very comfortable
- You are telepathic together
- You are not supposed to be together for a long period of time

Discernment

What is discernment and how do you learn to use it? Discernment is having insight, understanding and most importantly, good judgement. When you have had disappointments from your own judgement of situations and relationships, it makes you more cautious and more discerning. Discernment is the absolute key to recognising your beloved in person or in soul. The way that I recognise my Twin Flame in His soul presence is through:

- The intense vibration of love
- Poetry of love in His message
- Seeing his clothes, you rarely see his face for if you did you
- would not want to be in human form any longer
- A different more personal love than other guides, masters
- and angelic presences
- The feeling he is always there living within you and a part
- of your heart
- Absolute total protection emanating from the Twin Flame
- A feeling of wholeness
- A feeling of excitement and creativity
- No words to describe the bliss in His presence
- Discernment of His names and aspects over many years

Twin Flame Visualisation

There are many exercises and visualisations in Twelve Steps to Heaven to reach higher consciousness and your beloved Twin Flame. As in Twelve Steps to Heaven, please keep your chakras and higher chakras open to your Christed-Self and God-Self and clear these every day.

I am going to offer another simple visualisation to meet your beloved which can be done every day, whether your Twin Flame is in soul or in body. You are speaking to their higher-self which keeps your relationship pure and sweet.

Bridge to Happiness

Imagine a beautiful bridge and walk to the middle. You see a stream running underneath and you know its healing waters. You pour all your anxieties, worries and concerns into this water.

When ready, you walk over your Bridge to Happiness to your Garden of Bliss and you sit on a beautiful bench in your Garden awaiting your Twin Flame.

He/She comes to sit or stand next to you and greet you. There follows an embrace, words or a further journey with your beloved.

When you are complete you walk back over your Bridge and firmly back in your body.

TWIN FLAMES

Channelled by Azraella

Beloved Children of the Light - it is I Azraella. I speak to you as an aspect of Lady Jayne - I am in the Higher, Angelic. I work alongside my Beloved Lord Azrael, and we have been together since the beginning of time. We were caretakers and still are of the Angelic Realms.

Dear Ones, once you were together as one whole force, male and female but as you descended to more earthly existences, became separated and agreed with the Elders to do so. You went on a journey of your own, separate from your Beloved, and only maybe once or twice, came across your Beloved or even married your Beloved in a lifetime or two. But the plan was to lead lives separately; to bring Light separately to the World. Each of you have a certain mission; each very individual in that mission and as the lives progressed incarnation after incarnation, you found yourselves waivering, almost forgetting each other. Thus the darkness fell on many of us, many of you, and it took some lifetimes to clamber out of what you may term Hell to reach Heaven once more. And many of you are in your last incarnation and are about to or have met your Beloveds.

Your Beloveds may even be on the earth plane yet when two personalities come together there can be still some friction even though you know you are true Twin Flames at Soul. So one and the other have to be at the same level of evolvement to truly resonate, thus otherwise causing havoc in your emotional bodies. So in some ways, it is much purer and finer if one is on earth and the other in Soul. For the one in Soul already knows who she or he is in all their Aspects, for all their Aspects make up the One and mirror each of the Beloveds aspects who remain on the earth, dimmed from the memory, but are awakening to remember. Thus the Beloved in Soul draws them forward to remembrance. Eventually, you regain your sight, your bearings, and you regain your continued centredness feeling more and more complete, as your Twin beckons you forward, and in His or Her earthly incarnation that remains in Soul closer to the earth than the higher vibrations of Soul, He or She can almost walk the earth with Thee.

Thus, being seen by thy side; thus you envisaging their presence daily. Having more and more intimate conversations; having more and more laughter, having more and more Light pouring through your entire being. Thus, as the Light draws through you, it effects your physical body and yes, it changes the structure; the infra-structure, the DNA of your physical body. Thus, your body can learn to be in the Light and as your body becomes more and more of the Light you learn to disappear and reappear in an instant with thy Twins guidance. You are allowing thyself to travel the spiral of your being, of your souls essence to reach the Godhead. In the process of reaching the Godhead your inner soul's flame re-ignites and is seen by us and maybe by some in earthly embodiment. Your flame is activated above your crown reaching up to the top five at least of the chakras, above the crown reaching up to the Christed level and higher. And so you will regain clarity, you will regain divine purpose, joy and balance through your flame being re-ignited through your Crown upwards. For it is so used to being in the body traveling the earth doing mundane activities that are necessary while you are on the earth, yet it is forgetting during these mundane activities how to re-activate, until you take time to learn and feel your inner flame.

Dear Ones, you are on the verge of a GREAT discovery. You are on the verge of discovering who you are. You are on the verge of knowing who you are in the many aspects that you conjure up from your soul. These aspects work with the many Soul Beings that are awake, working for God. You are resonating to Soul Families in the many realms that exist in God's mansions and you will find your Home wherebeit, and you will know your Twin when you find this Home.

You will find many teachers on the earth are being activated to this higher soul awareness thus wherever you live or whatever part of the world you live you will find a teacher who will regain your memory of your Twins existence. And even if your twin is on the earth plane you can still re-adjust to his or her higher levels as you grow into yours, thus nothing is lost one way or the other - only to be gained; only Heaven on Earth to be gained. As you work with us in soul, you will remember that Angel part of thee, you will remember that Holy part of thee, you will remember the knowledge in one of your aspects, perhaps of Medicine, perhaps of Philosophy, perhaps of Science - whatever you care to remain in your Souls presences, <u>*will*</u> *remain. Yet you have the choice, but you do speak with the Elders to ask if these aspects are of sufficient benefit to the Soul family and others who may need this information.*

So dear ones, it is many levelled and very complex if you allow it to be so. I will not linger too much to explain the complexities of your Soul but I will linger on sharing the Love of your Soul. For the Love of your Soul, especially blended with your Beloved, is beyond human dreams, is beyond human language, for as Twins, we are made from the essence of God and for the most part is unfathomable in His/Her Love and so are you! For it is Eternal. You will be learning new and great wisdom eternally. You will be free as Twins to go as you desire to other galaxies, other planets, to understand all forms of being, all forms of light, all forms of intelligence, all forms of emotion, all based on Love. For Love is infinite, Love is beauty, Love is precious, Love is All.

For when Love truly ignites its inner flame to its highest point, and Love penetrates your subconscious mind into the body's awareness, into the lower chakras of the physical, you are ignited for this lifetime to travel in Light, in Love and in Service to the Supreme. When you feel this Love, you will know and you will know the Presence of your Beloved through this Love.

Call your Beloved daily every hour if you wish, for this gives Him/Her permission to step through for they do not want to come if you do not want their presence. It is all finely monitored so that you are not pushed in a hurry too forward and you take each step very slowly, subtly and gradually. Dear Ones, you are totally protected in this procedure for at this level there is nothing more than Love, Light and Support from the Elders, from the Masters, from God, from the Solar Logos, from all that have lived forever at this level and have watched over thee. You are protected and you are deeply loved and you are being drawn to where you belong and that is Home with a capital H!

I thee Love with all my Heart and all my being and I greet thee in. Thank you for listening to my words on Twin Flames.
Azraella.

One day my Twin Flame came to me and wrote this through my hands on the computer.

A Twin Flame Story

Thank you, thank you, thank you! For we are to begin our adventure! Long, long ago there was a beautiful princess and she had long golden locks and glorious fair skin. Her robes were of silk and gold and she glided into the room as if she had wings. I, the Prince, was so enamoured by her. We were blissfully in love and in adoration of every thought and every gesture. We were inseparable.

We were not possessive, nor jealous. We were in perfect harmony and unconditional in our approach. I wanted what would make her happy, and she wanted what would make me happy. We would often fantasise about travel to foreign lands and to have adventures together or apart. We always planned to come back to one another after our travels to share more delights. And this is what we always did. It was a promise between us. We would travel through the eyes of others too. We could do that because we were so Light. I watched the world through a Mongolian, an American Indian Chief, a Zulu warrior, a Rastafarian, a rich American Banker, a French Priest, an Astronomer, a Navy Captain and many more. My beloved Princess travelled too - she drifted from Sirius to Egypt, from Japan to China, to Bali and England. She had many English lives from medieval times to now. She always held on to her spirituality and chose delicate female orientated souls. I always chose souls that were forceful and had leadership qualities. Between our travels we would return to our dwelling on Sharkti, where we shared our journies. Sometimes she stayed behind and sometimes I did. This time I did. She needed the experience one last time and I said I would complete our journies here in my position in soul. I will evaluate our position and refrain from travel. I decided to travel through frequencies of higher dimensions and consolidate our position together. This is because this is her time to clear all our journies' emotional dross. She is clearing all mine and hers, as she is in the emotional field of Mother Earth. Once she has done this, and she is close now, she will be set free to reside in all our dimensions at once. I am waiting for my Princess to return and for her to say our work is done.

I have had many realisations in the higher dimensions. I am a child of God and all that that entails. I have the power that God bestowed upon His/Her children. I evaluate from my highest position and most powerful in terms of Love and Light where my energy should be distributed and how. Through the physical body of my Princess I charge her with this energy to place it correctly in ways her human mind does not comprehend. She becomes a conduit to apply Love and Light on Mother Earth and its inhabitants. Her human mind does not need to grasp this concept or the concept that Twin Flames can be or do anything within the Laws of the

Universe. They can set up Light Domes, Light Planets, Light Fields, Light Portals, and Light Universes if they so choose. I speak to all of those knowing or awaiting their Beloved Twin Flame, grasp the moment, take your power back to self and know your entitlement. More rapidly than ever before, we are entitled to flow from higher to lower in octave or frequency. Yet the lower dimensions do not mean dense, necessarily, just more of an experiential level of adjusting mental, emotional, and spiritual planes simultaneously. The vastness of knowledge acquired in one lifetime on earth will or could see you to a quicker ascension than that of a lifetime on any other planet in the universe! Your material world holds treasures of free will way beyond the comprehension of many. Thus, even when you are quiet and nothing seems to be happening, you are readjusting all levels of your being readying yourself for the next nugget of information, adventure, levels of ascension and so forth.

Adventures

One of my strangest adventures was when I became an Arcturian Leader and Warrior. Here we were trained to be a fighter to the end. We were relentless in keeping our empire pristine without any invaders from nearby planets. We were very proud of being Arcturians and very kind, but firm. In fact, we were rigid in our opinions and defiant in the upkeep of our Empire. So I stood a formidable being, defending my planet, unbeknown to my consciousness that my beloved Princess was waiting for my return. I did not return at the correct time and held off my return for several other lifetimes. My Princess, as unconditional as she was, had had lifetimes that had tampered with her purity and judgement. She felt abandoned, hurt and sad. We were now becoming conditional and found that to keep our original plan was becoming more and more challenging. To wipe the tears of my beloved, I knew our adventures had to continue and we would meet less and less. In fact our memory became almost obliterated through time. Still though, our higher selves were intact. It was only when we realised that our journies and adventures on other planets, including earth, were finally turning full circle could we register forgiveness and unconditionally once more. Perhaps to the human this was sad, yet to us, the experience was our ultimate saviour and we knew we could achieve our original position together when the time was right. And indeed there has been a healing of this memory with forgiveness. In In another adventure of mine, I was a great poet and writer of my day. I grew in notoriety and became the most talked about gay author in town! I was certainly the most camp and confident. Sarcasm ruled my wit, and decadence ruled my sensibility. I forgot what it was to be a Gentleman, rather a rogue of nuance and verb.

For me women were untouchable, beautiful and out of my reach. Yet in return for all my dallying, man spat in my face and flailed my heart. My experience thwarted by persecution, I returned to soul a broken man. This I knew to be my last dramatic appearance and I went out with style! Yet, defying all odds, my beloved Princess returned to me and whispered in my ear "I will return to earth to experience one more lifetime to clear our past and set each other free. This I do for you and for me, for to return to our sacredness is all that truly matters." And this she has done for me, my dear Princess. Another time, I was a French Monk. Indeed I loved to write the scriptures, day in and day out. I was in service so closely with God. I rejoiced at dawn and dusk in the name of God and His Angels. I vowed to be obedient and silent for many of those years. I was not allowed to think of my beloved, yet I knew in my heart there was a great love for me. I put this down to Divine Mother. I think I understood from that happy simple life, a sense of humility that I needed to learn, for so many lives were bold and courageous with male ego pride. I remember seeing a beautiful painting of an angel, she had long golden locks and she looked so familiar. Yet, who was she and how did I know her? I was not allowed to long for her or think of her. She just lived in my heart and her Love kept me content. I knew I would meet her again one day, I just knew it.

Another situation that I need to reveal to you is way before Lemuria we resided in another time-frame and galaxy. We were not human in appearance, nor did we have the human experience. You and I were one, neither male nor female. We resided mainly on a planet pronounced in your language Phylamandra. We were Light and Sound and pure unadulterated Love. We enjoyed long, as in hundreds of your years, lifetimes. Other beings from outer planets corrupted our peace and beauty and little survived of the Paradise we once lived in. We saw the rise and fall of communities and many destroyed. During this cycle you and I did descend to ascend. We held onto Twin Flame knowledge as we did not separate then. It was all a lot easier then to ascend; nor did we have human bodies and complexities to contend with of separation in the physical sense. We learnt to ascend as one. We are now learning to ascend as seemingly two. The richness of our existence on Phylamandra remains embedded in our DNA. Yet we have to realise we have never had such an experience of Ascension before and thus it is more challenging, yet not impossible. Consider this, most Twin Flame couples you are aware of are from this grouping and the beginning of all Creation.

In another life, I was a communicator on planet Saturn. I was always surrounded by masses of paperwork and I was constantly dictating news from many planets. Saturn was the main station for information and planetary news. We had a board of governors and I was a reporter at all meetings. I loved writing, so I was very happy in that role. Of course you probably wouldn't recognise me as I was very round and semi-reptilian. I had fingers, long ones and I could hold two pens in one hand. I needed to, so as to write two different sentences at the same time. Think of that! I suppose we thought of each other as looking beautiful, but from a human standpoint we would be a little strange to say the least. So I learnt my writing skills there and the ability to communicate succinctly and to the point. No humour though, I so missed the humour. Everyone was so formal and serious. I learnt discipline there to add to my repertoire of skills.

Did I ever tell you this one? I was wandering the earth wondering who could I be and what I could experience. I was drawn to China and became a Mandarin sage and soothsayer. Many people came to me for advice and direction in life. I helped people to find their true purpose and God. Well, one day the sweetest looking woman entered my workplace. She had jet black hair, straight and flowing down her back. She wore a pink satin dress. She bowed as I bowed to her. Her name meant "sweet music." My heart pounded for she reminded me of you my beloved. Memories flashed by as I knew her essence and her essence recognised mine. We spent many long hours together and she offered her beautiful singing voice to heal and calm others. She played the equivalent of a mandolin, so melodiously. She was a treasure. Do you remember beloved? Do you remember our happiness and reunion and the hours of laughter and joy? This experience was a reprieve, to rejoice in our loving energies in a lifetime on earth.

Another time in my higher aspect and between lives, I worked with Metatron as Lord Azrael, with hosts of angels. We, and those fighting angelic forces have soldiered through some dark energies that linger in the atmosphere of the grid work of earth. We continue to this day to release embedded dross that clouds the balance of earth and other planetary energies. Forces of darkness are rife in patches across the grids. As much as we build light, the darker energies persist. We are growing in numbers to activate the changes that now must take place on earth. We have called in all forces of light to be active in this metamorphic period of cleansing. We are the guardians of the light and let not fear disguise the truth that light will always win. For clarification purposes, there is indeed a conspiracy afoot in the modern western world yet despite the control that is being pursued, there will be hitches which will lead to a quick defeat. In the meantime the peacemakers will be strengthened to build the New World: A world of fairness and trust!

II

When we are enclosed in the physical framework and body of the human we remain locked into what we perceive as real. The reality is to be enjoyed for its beauty; and its beauty in all experience. The realisation that this is a dream made up by your higher-self and soul purpose, releases you from the illusion of life through human eyes. It releases you to value that which is truly real and that is that which lasts forever and a day - your soul and your union with your Twin Flame, God-Self, and ultimately, God Him/Herself. To fathom this is deliberately mystifying. Yet, when seeking those moments of reunion from your perspective seem so distant, so confusing, just close your eyes, quieten all thoughts and be in silence. Then listen within very hard, listen to your higher-self calling you, listen for your

Twin calling you and listen for God's words whispering love to you. Create a journal of your findings; allow your hands to just write and see what flows when the mind does not interfere, just as this hand is doing right now. Celebrate dear friends, for this moment is the only moment that matters. Tomorrow there will be more moments to celebrate. Fears from the past shadow tomorrows. You forget the present and the great joy of being alive, ever present in the moment. Each moment is a miracle of life flowing through you, helping you to create all that you desire and wish for, as you conjure up the wonder of life and gratitude for all that you are. For you know not who you are nor the individual soul that lies beneath the surface waiting for you to recognise your true nature. So I come back to my adventures.

Did I tell you this one? I was often peeking into others' lives and sensing the areas of existence I would like to experience. I found some wonderful people in the Himalayas and I chose them as my parents. I was brought up in the mountains far away from towns. We often visited a sacred area where the masters lived. I went there every week from a very young age. I was taught the ways of the masters. I was taught that whatever you think becomes a reality in some form. So I learnt to think of positive ways to improve my life and the lives of others. I walked with Saints, some of whom were hundreds of years old. You could almost see through them. They were real, but translucent. Sometimes they would disappear before my eyes and then return a few weeks, or months later. I often wondered where they went. I asked one day "where did you go Brother?" He was very old with a white beard and a loin cloth, with very kind eyes. "My son, I went to the Ashram in the sky. I brought back some bread and cheese to eat. Let us enjoy it together." He also brought back this sweet rice biscuit he knew I loved. "Here is your favourite," he would say. Out of nowhere the sweet rice biscuits would appear in his hand. "How did you do that?" I would wonder. "Ah, my son you will learn to do that one day, if you listen to the teachings and follow our advice." I lived there for hundreds of years. I became an old man to the young and I offered the same advice I once heard when I was young. I lived in the Ashram in the sky and visited my Himalayan grounds often. I would always be accompanied by other

Masters. I so admired them all and still do. We still can be heard if you listen carefully. I am your Master, beloved. I will clear the muddy waters if you were to just ask. I will show you how. I call you often beloved. Let me show the ways of the Master within. I will reveal all secrets when the illusion is surpassed into true reality, and the particles that hold flesh together part and all thoughts become infinite. Those things that sustain you, and form solid mass through the particles of thought projection, moving and then static, moving then static. Particles that blow in the wind, not tied together in the illusion. What you see before you is a belief of solid. The solids are particles that blow in the wind. We are many forms. Through thought and particles we become those forms. In one breath we can disappear and become another form. We can play with colour, light and sound and weave our magic. We are magicians. You and I beloved. Remember?

III

Watching the stars at night I think of you. I gaze at the universe and galaxies and I wonder where you are and what you are doing and thinking. What is it that you are needing to achieve right now? And I look into my heart and I know all these answers and more. There are no more questions or unfinished stories. There is only our reunion. I long for your company in your wholeness and yet I have it here in my heart. You see beloved, all your longing and yearning is mine and vice versa. Yet, when the time comes so close to completion, all that was so challenging, disappointing, sad, and heart-wrenching, disappears in a moment. And Here We Are. A new paradigm, a new way, a new future together. I watched you once. You were sitting in a field, on the long grass, sketching the autumn day. And I watched you depict the beauty of nature and I whispered to you. "Not so far away are you from the real dream that beckons you forth. Meet me just half way and I will show you wonders beyond your imagination. Come dear one." But you felt me and did not hear. You were happy in that reality. And so I stood back to rejoice in your happiness and not interfere. You see my dear, I have waited eons for you to hear me again and celebrate our remembrance of one another. It was laying deep in your heart of hearts ready to awaken. And now of course you are that beautiful Princess that has awoken after a million years of sleep. Still we know we did what we had to do for it was our promise to God and ourselves to serve in any which way we chose. Yet, the promise is fulfilled and thus, now, we are writing our story together before we move on to a place where our story does not matter. Where we start again anew, yet, free to choose our destiny and creativity together. We could build new realms for those souls who need to reach contentment and peace. We could build castles and fjords if we so desire. We could build love clouds for angels to rest on. There are endless possibilities beloved. I await thee.

IV

So it is, my beautiful Princess. We still meet as Prince and Princess and we still flourish in each others love. We are planning all the time for new adventures and how to exquisitely relate to the physical body to relinquish its hold on thee whilst not disappearing completely from the human realms. It is a delicate pursuit we follow indeed. Sometimes I see you rock with fear and trepidation yet the next day I see you fully resurface understanding that fears lying in our deepest crevice must be revealed and released. We are indeed, fighting in a comfortable way, for our right to be free. Every inch is another inch closer to our departure from the ways that we once knew. It is all so exciting as we acknowledge our strength and weaknesses yet know that our soul is All and especially together. And so it is.

Deliberately the mind is veiled from so much that goes on in the background. Twin Flames and other guides are working diligently to allow more information to come to the surface of those who are ready to take the leap of faith into Ascension and movements towards channelling their own guidance. It is painstakingly designed to allow those guides on this side to interweave with those on the human side so to speak. Heightening your frequency as we lower ours. We meet somewhere in the middle through a very fine filter of consciousness attuned to a certain wave pattern. To catch this wave, your mind needs to learn to be still and alert. Fine-tuning is taking place for many now even those who are seemingly asleep are impressed to re-tune their instrument.

V

As we await our true position and a lengthy process it must seem to you, we are realigning our aspects to one another in the higher realms. Each one is seen as a renewed marriage as to practice true consistency of energy and alignment. Even though you are not sure which aspects are aligned and which are not, I am making sure they are all aligned. You are already feeling that these aspects are within you now. And that is so. There will be a feeling of no separation and that is when you know the job is done. You have perceived for a long time the coming and going of my presence. Now you will perceive me as not being separate. It is true to say that as a human in your era on earth, the trappings of the consumer world are enticing and the needs of the flesh need to be met in terms of food, clothes and shelter. Yet, these activities are a lesser form of need of your spirit and soul. The soul needs clothes in the form of rays of colours. It needs food in the form of Love and it needs shelter in the realms of heaven. Choosing the beauty of the world to surround you is indeed a reflection of your soul's needs. For the colour of nature will feed you, Mother Earth's Love will nurture you, and the sun will nourish you. Living as intended on God's earth will bring you to a place of harmony within. More and more you as a society, are turning to the simple pleasures of life. For the concrete surrounds and the hustle bustle of city life cannot ultimately grant you your needs or your comfort. Rejoice each morn with the song of the bird, and retire with the whisper of my Love.

VI

As from my view and perception, that which you consider human frailty is fear based and not in your DNA but in your subconscious and this is easier to clear. Your DNA is perfected now. All you have to do is wait for more of your talents, and light body comes to the fore. This may take weeks or months. I know you are patient beloved and your patience has paid off. The final layers are being peeled away to share your shiny divine self to the world, permanently ungrazed by the human experience. You will be a shining example of how to Ascend and you already are. We are now resting in the final branches of perfection. As you well know, when the time is right you will be shown the next stage of flowering with all your petals awakened and blossoming, your entire being freed from negative memory and false limitations and beliefs. All that is needed from you is to allow and go with the flow of expansion that is occurring for you and through you. We will walk in all dimensions hand in hand offering pellets of wisdom and knowledge to those who wish to receive. Worry not dear one, they indeed will come.

Chapter Three

PATTERNS AND PROGRAMMES

We all carry patterns and programmes from this life and other lives, from both our childhood and adulthood. To have a healthier and happier life, we can choose to work on releasing and healing our past. There are many therapies and techniques of your choice that can help this process. I have personally received many therapies over the years, including Spiritual Healing, Reiki, Coaching, NLP (Neural Linguistic Programming), Hypnotherapy, Channelling, DNA Activations, and Akashic Record Clearing.

It doesn't matter how much work we have done to clear both our negative subconscious and conscious patterns and habits, there will always be something to work on. It does get easier over time, that is for sure. As we progress in our journey, we need to have light moments and to absolutely remember to laugh during the work we do on ourselves.

If you don't already use it, one exercise that is an absolute gem (to use for life) for clearing relationships that drain us in any way or that are unhealthy, is Cutting the Ties - originating from books by Phyllis Krystal written in the nineties.

Cutting the Ties Meditation

Close your eyes and take three deep breaths.

Now imagine yourself somewhere quiet on a beach where there are no people to distract you. You feel a sense of freedom and connection to nature, the sun, sea and sand. Now imagine yourself drawing a figure of eight on the sand and you stand in the centre of one circle. You call the person you are the cutting the ties with into the other circle. You thank the person for coming. You say everything to them you have never said, whatever that may be. Then, when you are complete, you find a pair of scissors and cut the ropes of energy that have formed between your chakras. You cut the rope at the crown chakra from theirs to yours, completely, and start to form a pile of these ropes a little way away from you. Then you do the same for the chakras at the third eye, throat, heart, solar plexus, abdomen, and base of spine. If you feel you need to, you can also cut the ropes from the knees and ankles between you. Once all ropes are cut and are sitting on the pile, get a match and burn all ropes. See the ashes fly away in the wind.

You then place the person in a bubble of unconditional love and say all of these statements three times,

"I forgive you", "I accept you" and "I love you unconditionally."

Thank the person for coming and see them float away on to their own chosen karmic path. Then place your yourself in a bubble of unconditional love and say all of these statements three times,

"I forgive myself", "I accept myself" and "I love myself unconditionally.

Slowly come back into your body and open your eyes.

Some programmes seem to recur repeatedly and are deeply embedded in our psyche, despite years of work on self. Do not be disappointed, or discouraged in yourself if this happens. Please keep doing the work necessary for your own clearing. Above all, keep Loving Your Self during the process of peeling away the layers of programmes.

Inevitably, the most recurring areas to clear are your mother and father issues from this current lifetime, plus any of their shadow side(s) that you are carrying. If you truly desire to evolve, it is imperative to heal your childhood issues, your inner-child and your own shadow side. Past life emotions are also issues that appear over and over again.

Some of a multitude of techniques for clearing are working with:

- St. Germain's Violet Flame,
- Archangel Michael's Sword of Light
- Removal of axes, knives, or spears representing anger and other emotions
- Dissolving energetic calcified eggs that hold difficult memories
- Removal of karmic patterns living behind the spinal column
- Cancelling all negative beliefs from all timelines
- Cancelling of all vows, contracts and promises which can be accomplished by turning geometries of emotions in the body into orbs

Ascension Work includes clearing and releasing of all issues with:

- Mother
- Father
- Siblings
- Childhood
- Inner Child
- Partnerships/Relationships
- Past Lives Trauma, Fears and Negativity
- Current Life Trauma, Fears and Negativity
- Twin Flame - Past Lives held in your field
- Creator God Misunderstandings
- Negative Subconscious Memories
- Negative Body Memories
- Negative Cellular and DNA Memories
- Karmic resolution with soul mates and all others
- Inherited Family Issues in the DNA
- Carrying others' karma and emotions as well as your own

Working with the DNA and Akashic Records takes you to ever deeper levels. You may find redundancies of the same stories you have already worked on and even after doing so, another core chapter emerges to be resolved. At this point one can be disappointed that there is more to release, but it is worth persevering. Learning to accept and love yourself, wherever you are in your self-development, is very important and also, above all, to forgive yourself and your patterns. Thus one becomes more tolerant of others and can more easily forgive them for their shortcomings. The stories stored in our subconscious, akashic record computer, DNA, cells, muscles and body need to be released to allow your life journey to be more peaceful and joyous. We have to take each step and not miss any to reach the ultimate goal of becoming a clear channel of Light and Love, whether we are a healer, intuitive or someone yet to discover these, or other talents. Basically, the result of becoming a clear channel is to set ourselves free and to Ascend.

With all our efforts we reap the rewards and moments of bliss and serenity. We find more access to our higher-self, soul group, angels, guides, Twin Flame and Mother/Father God. We have more moments of remembering who we truly are and recognise more of our soul. We could be fortunate enough to see our own rainbow light, and know our Divine Wisdom and the Divine Love that we are. We create moments of reunion with the Truth more frequently than ever before and the New Earth vibrations set us free to do so.

Other work to do on self:

- Reuniting Twin Flame Higher Aspects with Your Twin Higher Aspects
- Recognition of Divine Love with your Twin Flame
- Twin Flame higher reunion in the physical body
- Meeting Soul Group in Higher Dimension
- Male and Female Balance in the physcial body
- Bringing all higher energies into the physical body

Physical Symptoms can include:

- Back Ache
- Sciatica

- Shoulder Ache
- Headache
- Sleeplessness
- Tiredness
- Disruptive Dreams
- Muscles Spasms
- Pinched Nerves
- Having cold/flu symptoms - which is an emotional release
- Ultra sensitivity to foods, drinks, and chemicals in foods
- Hot flashes and night sweats
- Inability to mentally focus
- Loss of memory
- Indigestion, stomach aches, diarrhea
- Nausea and vomiting
- Body twitches and jumps
- Skin eruptions and pimples
- Hot Feet

Some Ascension Symptoms:

- You have lost your desire to be a healer
- You feel that your life is "here" and not back "home" somewhere
- You give up any desires to save anyone or anything
- You have stopped belonging to groups and discussing spiritual matters
- Your personal energy is more "subtle" than it used to be
- You feel invisible when you go out in public

- You lack for nothing
- You set up new boundaries with friends and people in general
- You put yourself "first" in regard to personal health and happiness
- You are comfortable refusing to participate in things that feel uncomfortable to you
- You feel more peace and tranquility
- You find all your answers within rarely asking for outside help
- You know you are not perfect and never will be
- Faster processing of old patterns that need to leave
- Flip flopping from the old you, to the new you
- New friends and loss of some old friends
- Your home is the only energetic sanctuary

Daily Affirmations for Your Ascension and Releasing of All Vows

I now call forth Mother Father God, my guides, masters, angels, soul group and Twin Flame.

I now release and clear all remnants of my karmic debts, patterns and programmes. I release and clear all conditions easily and effortlessly.

I now release and clear all vows and contracts from this life and all lifetimes.

I now release and clear all lifetime memories where I renounced love.

I now release and clear all agreements with my mother and father, family members and friends that keep me caught in the third dimension.

I now release and clear any need to save the world or anyone in it and I realise my mission is to accept my own self mastery and be a living, loving example to people around me.

I now release and clear all judgements about other people, knowing that they are in their own perfect place and evolvement.

Male and Female Balance in the Body

If you happen to feel that you are unable to meet your Twin Flame in His/Her higher soul level, please note that for your ascension it is important to balance your male and female energies within. This is vital if you wish to grow spiritually in perfect balance. It is also necessary to feel the male/female balance in the chakras, body and cellular level. Many times, spiritual and creative people find it very hard to work with the lower chakras that deal with third dimensional life and especially the root chakra, which is linked to finances and stability. The lower chakras are linked to physical life and are associated with the male aspects. These chakras need to be nourished more, especially if there are weaknesses with living in the third dimension.

We are not trying to escape physicality, but to be part of all dimensions, therefore we are ascending with our body and not without it. In previous times people ascended leaving the body behind. We are in a new way of being that has never been realised before as humans. Yet, instinctually, we know how to do it and how to follow our intuition during this process of growth.

Balancing Male and Female Exercise

Relax and close your eyes.

Call your female aspects to spiral down your left side and the call your male aspects to spiral down your right side. Ask for them to be in balance and in harmony.

Then merge the two spirals together - One up and the other down your spine. Slowly open your eyes when you are ready.

From this exercise, you will realise if there is any dis-harmony between male/female aspects and if so, I suggest you go to a competent healer or therapist that knows how to assist in releasing any blocks or negativity still held there.

Chapter Four

CHANNELLING

Especially in the USA, channelling is a very popular form of talking to higher beings such as angels and masters of wisdom, insight and love. Many people are becoming channels more than ever on different levels, utilising either their solar plexus, third eye, higher self, guides, angels or soul group. There are many who are channelling well known guides, such as Archangel Michael, but do not always understand how to reach his true soul level. People also do not understand they can be channelling through their own ego levels and a lower level of the core soul of Archangel Michael.

To dispel much confusion and delusion of the ego, I have put together a list of channelling levels for your discernment.

Levels of Channelling
by Azraelus

Stage One

Base Ego - Kindergarten levels up to about stage three
Signs:

- early psychic awareness,
- cards, psychometry etc.
- utilising solar plexus, lower ego-self, base-astral, self-opinionated messages
- dogmatic, subconscious interference, conditional, not healing

Stage Two
Ego/Conditional

- some basic mediumistic skills
- awareness of spirit guides e.g. - american indian, chinese man, etc
- ego-self of guides, base-astral, self-opinionated messages
- drawn to be a healer
- subconscious interference

Stage Three
Ego/Conditional and Unconditional

- complete awareness of spirit guides
- working as a medium/trance medium
- spiritual church activities
- proof of life after death, awareness of family members who have passed on
- clearing some past lives
- intermediate ego-self, self-opinionated messages
- subconscious interference
- desire to become a healer

Stage Four
Ego/ Higher Self/Loving
Intermediate Level

- awareness of higher vibrations of angels, masters etc.
- clearing past levels
- messages from masters
- spirit guides and merging guides with masters, acknowledging false masters, inconsistent messages, mundane psychic work, some lower ego-self, intermediate ego-self, higher ego-self, some higher self, some soul work
- still dealing with stage one to three at times, unconditional love growing stronger, becoming a full strong healer,

becoming fully self aware to clear all blocks, self honesty medium to high

Stage Five
Surrender of Ego

- Divine Love/finely tuned subconscious/ Higher Self/clarity/ more unconditional
- pure soul, soul family/group awareness, consistent flow of channelling, ascension work, spiritual psychology, wisdom, residue of past life, this life work, masters, angels, Twin Flame reunion, Christed Self, God Self awareness, higher self, purity,
- purification of stages one to four use in some ways, little future or mundane messages, healer of all levels of being, little interference of subconscious if any

Stage Six
Master 1st Stage

- choice to channel or not, Twin Flame unified, Christed Self apparent, full ascension awareness, hooray!

I cannot stress enough of one's need to be discerning when you go to a channel for a reading or develop as a channel yourself. It is essential to make sure the absolute highest and purest energy, Angels and Guides are called through. If you have any doubts listen to your intuition. There are a lot of people who are primarily working with their ego and people can be very vulnerable in these situations. Please use your discernment.

To work with those on high, please call firstly Mother Father God and the Risen Christ, Lord Sananda and surround yourself in their Love and Light. Then ask to work with the highest soul level of the Masters, Archangels, Angels, Guides and your personal Soul Group to come through. Pray every day in this way to develop to your highest potential. Please note that God and your own guidance know what is best for your vibration and the role you are to play.

Not everyone is destined to be a professional channel, yet there may be other gifts you can channel such as healing, writing, art, or other creative expressions. Becoming a healer is a another special way to open your channel to messages.

Tips in Becoming a Channel of Love and Light

- You need to be pure of heart, of integrity, happy to help others, always willing to work on your own issues and ready to take the responsibility to be a channel for others.

- Becoming a channel is a part of a higher spiritual agreement usually made lifetimes ago. We have all chosen different tasks on earth, so we do that which makes our heart rejoice. Be cognizant to check at a heart level if it feels right to be a channel and not an ego or mind decision.

- Self-honesty regarding your own negative patterns and habits is vital for your healing process and also understanding how to help others. Clearing your subconscious mind of this life's issues along with past lives' issues allows you to be a more pure and clear channel, healer and creative being.

- You need to be balanced and centred allowing your Guidance to be in control, yet not to the extent they take over your life where messages are given anywhere and everywhere.There is discipline in knowing how the messages are to be relayed at the appropriate time and place. At first one can become obsessed with the whole process, so remember to keep your feet on the ground and find an excellent experienced teacher.

- Your ego, emotions and mind need to step aside and know that "Thy Will Be Done" and not your will. Meditation helps you to calm the mind and practice peace.

- You need to surrender a part of your heart to love and service, with absolute pure intentions. Your pure vibration will draw in pure guidance and messages. Have high standards and keep watching the ego creeping into the messages. Remember you are in control and you will ask for the best repeatedly.

- When names are given of your guides realise your guides could change over time as you grow and develop. Some people have the same one guide their entire life, others a handful, and others a whole soul group.

- When names are given that are famous, whether that be of your guides or Twin Flame, keep checking over a period of years that this is consistent and true. Do not jump to conclusions of the famous or remember them as they were on the earth plane. They will come in at a higher level than during their human life. It is also very common to remember you were famous in a past life. This more often than not is an archetype attachment to the story of someone famous and not that you were really that person. Logically most people are not famous in this life or others.

- The message should be self-empowering and uplifting and not fearful or being annoyed and angry with you. There are ways to be inspiring and caring about matters that need to shift or change.

- The best form of channelling will never, ever channel a message that asks you to give up your free will.

- Also, the messages should not violate the integrity of what you believe. You need to feel comfortable with the information given, and it will ring true to your heart.

- Channelling from a good source has no ego and is of a love and sharing that is for all people.

- Information coming through should change, grow and be new and offer solutions to life challenges.

- Some channelling is for groups only and impersonal, and other kinds are for a one on one personal level. It depends on the channel's role that has been chosen on higher levels.

Dr Kruger on Becoming a Channel

Beloveds,

There is a misunderstanding often in groups who are meditating together and bringing through spirits. Not all spirits are of a high calibre or of integrity and it should not be presumed if its a spirit it knows better than you!

There is also a misunderstanding and belief that bringing loved ones through should be the be all and end all, nor should the process be supported if the channel or medium is imbalanced emotionally. The emotional filter is not a solid proven one and thus the messages relayed through this imbalanced person can be misinterpreted very easily by the person. Not only that, the person can be very susceptible to those spirits who would wish to play on the emotions of others.

Do not presume ever that the spirit speaking is who they say they are, nor presume that they are coming from a place of Love and Wisdom.

The measuring stick is in your hearts intuition. If the message is full of "you have to do that or do this," stating the future is full of bad things and infer that you only listen to them and no-one else, should ring alarm bells to their authenticity.

Remember you have free will dear ones. No spirit or higher soul beings have any right to tell you what to do and demand their way is the better one. Plus, your free will can create a better tomorrow always, as you are the one designing your day and your responses to life.

If the messenger through a channel is consistent in Divine

Love, and is caring and compassionate, then you know you are on the right track.

From spirits perspective we have to be trained just like you, to be able to be a guide from this side of the veil. We can train many of your years to do so. Then we have to find the right person of integrity, ethics and open heartedness, and of the same vibration as the guide, to even start the process of partnering. Right partnering is crucial for us otherwise our own spiritual growth can be put in jeopardy. It has to be a perfect match from us to you and you to us.

As soul beings we would prefer you to be more concerned with others than your self, to have proven an earnest desire to be a spiritual helper, and to know your limitations and be open to rise above them, keeping a sense of balance and humour!

Beloveds it may seem very glamourous to be a channel, but you have a huge responsibility not only to yourself and us, but to others. We would also like to add that you do not have to be a channel to be a great healer. A great healer is just as important as are your spiritual creative gifts, which may have a greater impact on more people and the world at large, than you realise.

Ultimately we wish you to channel your greatest potential, thus leading you to your true fulfillment.

God bless you on your unique path of self-discovery in communication with the spiritual realms.

I am always in service, Dr Kruger.

We all have been given the gift of time and energy. When we give this gift back to God by dedicating our time and energy to finding out who we are and living our life purpose, the Universe, in turn, responds by supporting us in this search. As we keep practicing, we develop a force field around us that magnetises and energises the process. The time and energy we dedicate to this search of our true life purpose is magnified many times by God, so that growth for each of us becomes almost effortless.

During each day, stay in alignment with your Higher Self and guidance. Ask to create a day full of love, joy and purpose. Ask how best to serve people, have peace time, send absent healing prayers for those in need and try to stay and be in the moment.

Prayer for an Ascended Master Relationship

Dear Mother-Father God, Lord Sananda the Risen Christ, All Masters, Archangels, Angels, My Soul Group and Twin Flame.

I ask and pray with all of my being, heart and soul for your combined help to magnetise and attract to myself an integrated Ascended Master Relationship to assist me in establishing constantly for myself, an Ascended Master, fully embodying Christ Consciousness, Enlightenment, Wisdom, Divine Love, total radiant health with the elimination of negative ego to the highest potential.

I ask for your combined help to let go of any negative ego or attachment associated with any patterns or programmes held in my subconscious mind, Akashic Records, DNA and cellular memory.

Please Mother-Father God, I am truly choosing the mantle of the Risen Christ and the release of the Old Matrix mass consciousness in favour of integrated Christ Consciousness.

Help me to prepare for this gift of Heavenly realms in humility and total appreciation, so that it serves for the Greater Good of All.

I hereby ask and command that you take this prayer, with all the vital force that is needed, to manifest and demonstrate this prayer to the source of my being through my beloved Soul. Amen

Chapter Five

DNA ACTIVATIONS

After receiving DNA Activations, studying and reading much material on DNA, I have devised and channelled Eight Levels of DNA Activations thus far, which range from twelve strands to multi strands. It has been an ongoing process and discovery since 2004 for me.

DNA, or deoxyribonucleic acid, is the hereditary material in humans and almost all other organisms. The information in DNA is stored as a code made up of four chemical bases: adenine (A), guanine (G), cytosine (C), and thymine (T). Human DNA consists of about 3 billion bases, and more than 99 percent of those bases are the same in all people. The order, or sequence, of these bases determines the information available for building and maintaining an organism, similar to the way in which letters of the alphabet appear in a certain order to form words and sentences.

DNA bases pair up with each other, A with T and C with G, to form units called base pairs. Each base is also attached to a sugar molecule and a phosphate molecule. Together, a base, sugar, and phosphate are called a nucleotide. Nucleotides are arranged in two long strands that form a spiral called a double helix. The structure of the double helix is somewhat like a ladder, with the base pairs forming the ladder's rungs and the sugar and phosphate molecules forming the vertical sidepieces of the ladder.

An important property of DNA is that it can replicate, or make copies of itself. Each strand of DNA in the double helix can serve as a pattern for duplicating the sequence of bases. This is critical when cells divide because each new cell needs to have an exact copy of the DNA present in the old cell.

At most, we are only using approximately 10% of our DNA or our brain. The DNA Activations I perform work with the 90% of the DNA not being utilized. This unused 90% holds our latent talents and spiritual gifts, as well as enhancing divine youth and perfect health. I use a good deal of sound work throughout the activation levels, as well as the spoken word and through these give specific commands to the Master DNA Cell within the Pineal Gland. The instructions are then replicated in the trillions cells in the body.

The first DNA levels, I and II, work with the first twelve strands. Science generally recognises only two strands of DNA and regards the rest as "junk DNA." The so called junk is esoterically believed to be a link with interdimensional and cosmic wisdom, which needs to be reactivated in all of your cells. In subsequent DNA levels, I work with twenty four strands progressing to even higher strands on multi-dimensional levels. As humans, we have no comprehension of the capacity of our DNA, or our latent talents, which include teleportation, manifestation, x-ray vision, and so much more. DNA is the blueprint for your life purpose and divine potentialities - for who you are as a Divine Being. It is God-given, holy, and sacred. It defines the uniqueness of you.

DNA Activation is a verbal and quantum energetic initiation that triggers some of the dormant 90% that is not yet being used and can include:

- Awakening your latent potentials and true power
- Increasing your extrasensory perceptions and intuitive faculties
- Clearing emotional blocks, addictive and habitual patterns
- Stimulating past-life memories for clearing and reawakening dormant talents and creativity
- Enhancing your clairvoyant, clairaudient and clairsentient abilities
- Balancing your physical health and well being and harmonising your bio-energy field to increase levels of peace and bliss
- Increasing your ability to maintain your connection to Source for prolonged creative expression

- Connecting you with many levels of Universal wisdom and your path to Ascension
- Empowering your Path of Light by releasing coded information in your DNA that allows you to clearly see your life's purpose and mission

DNA and Science

It has been reported by Russian Scientists Von Grazyna Fosar and Franz Bludorf, that human DNA is a biological Internet and superior in many aspects to the artificial one. The latest Russian scientific research, directly or indirectly, explains phenomena such as clairvoyance, intuition, spontaneous and remote acts of healing, self-healing, affirmation techniques, unusual light/auras around people (namely spiritual masters), mind's influence on weather patterns and much more.

In addition, there is evidence for a whole new type of medicine in which DNA can be influenced and reprogrammed by words and frequencies without physically cutting out and replacing single genes. Only 10% of our DNA is being used for building proteins. It is this subset of DNA that is of interest to western researchers and is being examined and categorised.

The Russian researchers, however, convinced that nature was not dumb, joined linguists and geneticists in a venture to explore that 90% of "junk DNA." Their results, findings and conclusions are simply revolutionary!

According to their findings, our DNA is not only responsible for the construction of our body, but also serves as data storage and communication. The Russian linguists found that the genetic code - especially in the apparent "useless" 90% - follows the same rules as all our human languages.

To this end they compared the rules of syntax (the way in which words are put together to form phrases and sentences), semantics (the study of meaning in language forms) and the basic rules of grammar. They found that the alkaline of our DNA follows a regular grammar structure and do have set rules, just like our languages. Therefore, human languages did not appear coincidentally, but are a reflection of our inherent DNA. The Russian biophysicist and molecular biologist Pjotr Garjajev and his colleagues also explored the vibrational behaviour of DNA. In brief the bottom line was: "Living chromosomes function just like a holographic computer using endogenous DNA laser radiation." This means that they managed, for example, to modulate certain frequency patterns (sound) onto a laser-like ray, which influenced DNA frequency and thus the genetic information itself.

Since the basic structure of DNA-alkaline pairs and of language (as explained earlier) is of the same structure, no DNA decoding is necessary. One can simply use words and sentences of the human language. This, too, was experimentally proven!

Living DNA substance (in living tissue, not in vitro) will always react to language-modulated laser rays and even to radio waves, if the proper frequencies (sound) are being used. This finally and scientifically explains why affirmations, hypnosis and the like can have such strong effects on humans and their bodies. It is entirely normal and natural for our DNA to react to language!

While western researchers cut single genes from DNA strands and insert them elsewhere, the Russians enthusiastically created devices that influence cellular metabolism through modulated radio and light frequencies, thus repairing genetic defects.

They even captured information patterns of a particular DNA and transmitted it onto another, thus reprogramming cells to another genome. So, they were able to successfully transform, for example, frog embryos to salamander embryos simply by transmitting the DNA information patterns! This way the entire information was transmitted without any of the side effects or disharmonies encountered when cutting out and re-introducing single genes from the DNA This represents an unbelievable, world-transforming revolution and sensation: by simply applying vibration (sound frequencies) and language instead of the archaic cutting-out procedure, DNA can be reprogrammed to other frequencies, which can literally transform one form into another.

This experiment points to the immense power of wave genetics, which obviously has a greater influence on the formation of organisms than the biochemical processes of alkaline sequences.

Esoteric and spiritual teachers have known for ages that our body is programmable by language, words and thought. This has now been scientifically proven and explained. Of course the frequency has to be correct and this is why not everybody is equally successful, or can always do it with the same strength. The individual person must work on the inner processes and development in order to establish a conscious communication with the DNA.

The Russian researchers work on a method that is not dependent on these factors, but will always work, provided one uses the correct frequency. Most importantly, the higher developed an individual's consciousness is, the less need is there for any type of device.

The Russian scientists also found out that our DNA can cause disturbing patterns in a vacuum, thus producing magnetised wormholes. Wormholes are the microscopic equivalents of the so-called Einstein-Rosen bridges in the vicinity of black holes (left by burned-out stars.)

These are tunnel connections between entirely different areas in the universe through which information can be transmitted outside of space and time. The DNA attracts these bits of information and passes them on to our consciousness. This process of hyper-communication (telepathy, channelling) is most effective in a state of relaxation.

Stress, worry or a hyperactive intellect, prevent successful hyper-communication and the information gained will be totally distorted and useless. In nature, hyper-communication has been successfully applied for millions of years. The organised flow of life in insects proves this dramatically. Modern man knows this on a much more subtle level as "intuition." But we, too, can regain full use of it.

In humans, hyper-communication is most often encountered when one suddenly gains access to information that is outside one's knowledge base. Such hyper-communication is then experienced as inspiration or intuition (also in trance channelling). When hyper-communication occurs, one can observe in the DNA, as well as in the human, supernatural phenomena. In order to develop and experience individuality, however, we humans had to forget hyper-communication almost completely.

Now that we are fairly stable in our individual consciousness, we can create a new form of group consciousness - namely one in which we attain access to all information via our DNA without being forced or remotely controlled regarding what to do with that information. We now know that just as we use the internet, our DNA can feed proper data into the network, can retrieve data from the network, and can establish contact with other participants in the network. Remote healing, telepathy or “remote sensing” about the state of another can thus be explained. Some animals know from afar when their owners plan to return home. This can be freshly interpreted and explained via the concepts of group consciousness and hyper-communication.

Researchers think that if humans with full individuality would regain group consciousness, they would have a god-like power to create, alter and shape things on Earth and that humanity is collectively moving toward such a group consciousness of the new kind. When a great number of people become unified with higher intention, as in meditating on peace, potentials of violence also dissolve.

DNA is a living, vibrating holographic model of the Universe contained within each cell of the human body. It is one of the ways we communicate inter-dimensionally with All That Is and has all potentials within it. Human DNA has not been utilized to its highest potentials, but can, through proper sound and language activation, be actuated so that we can progress into the next stage of human evolution.

The Sixty Four Codons

The genetic code, comprised of codons, is the set of rules by which information encoded in genetic material is translated into proteins or amino acid sequences by living cells. The code defines a mapping between the tri-nucleotide sequences and the amino acids.

It is said there are Sixty Four codons to generate the amino acids of the double helix of the DNA structure. Out of all the combinations only twenty out of the Sixty Four codons generate the DNA amino acids. In my own discoveries in the DNA Activation process, I am able to activate Four more codons in addition to the Twenty Codons already in use. Indigo Children and New Earth Children are said to already have the extra codons activated.

During the Activation process, I intuitively count the Sixty Four Codons of a person, and find I receive completely different information to the feedback of the DNA Strands. From the Sixty Four Codons, I receive the innate talents of the person in this incarnation that are able to be utilised and activated.

Information from the book *Twenty Tablets of the Law of Time by* José Argüelles and Lloydine Argüelles is the culmination of their life work. The foundation of the information contained in the *Twenty Tablets* is the discovery of the correlation of Sixty Four rune symbols to the Sixty Four hexagrams of the I Ching. This information is carried to another level by the realisation that these Sixty Four runes and hexagrams also correlate to the Sixty Four codons of DNA. The discovery of the correlation of the runes, hexagrams and DNA brought through a series of meditations and telepathic exercises which will allow us to reset and activate our DNA patterns to the level of solar-galactic consciousness.

The connection of the rune symbol to a hexagram and a DNA codon led to another discovery. That discovery was that if you multiply the energy of the rune symbol by the thirteen ancient galactic tones of creation, you carry the rune symbol, the hexagram and the DNA codon through all of their possible permutations. The result of which is that our DNA is reset to its natural state and activated.

According to Nassim Haramein, (www.theresonanceproject.org) Sixty Four is part of a code in the geometry of all life and creation. The Star of David and the Kabalistic Tree of Life of the Hebrew tradition, the Flower of Life, the Taoist Sixty Four grid of reality, the Yin and Yang sign, Sacred Geometry, the Egyptian Eye, the Mayan Gliffs, the Pyramids, and other religious symbols around the world, specifically describe the same geometry of creation. Everywhere you look in ancient traditions describing the God Force, you will find allusions to the geometry of space-time embedded as a code that could not be deciphered until we reached the level of understanding that we have today! Eight Kabalistic Trees of life comes together to form a Sixty Four tetrahedral grid. In fact, the translation of the word Yahweh, the Hebrew name for God, is "Tetra Graviton" referring to four sides of a tetrahedron and perhaps gravitation.

The Flower of Life is a representation of a wave form when the Sixty Four tetrahedral grid is seen in three dimensions. The yin and yang sign is a representation of the spinning double torus as seen from the top. The I Ching is based on a Sixty Four grid of reality and the Star of David is the geometric shape used to construct a Sixty Four tetrahedral grid in 2D.

When the negative space was accounted for inside a tetrahedron, it formed an octahedron, which Nassim thought was the blueprint of the universe, but it wasn't polarised so he knew there must be two needed to fit the equation. When this was put into a sphere and pushed together, it made the Star of David inside a circle. The central geometry is the vector equilibrium. Octahedrons are conjoined pyramids with another one below it.

Vector equilibrium is where all points are equal distance apart on the perimeter, as well as from the center (a vector is a quantity with direction and magnitude). A tetrahedron has twelve vectors radiating from the center equilibrium and twenty four total. The numbers twelve and twenty four are found in most ancient texts. When you add tetrahedrons, you get a total of Sixty Four vectors, the same number as our DNA structure.

Many crop circles have very specific geometric shapes relating directly to the key to understanding the nature of creation and the Sixty Four tetrahedral grids that infuses the vacuum.

Sacred Geometry and the DNA

Sacred Geometry is inherent within DNA and with life itself. There has been some interesting research regarding the two, including how they are significant in the vitality of water and life. One scientist, Dr. Fritz Popp, a bio-physicist, discovered a most important function highly organised water plays in the body. He found that our DNA codes are transmitted to messenger RNA by the emission of mini ultraviolet light bursts he called "Bio-Photons." The mechanism for this ultraviolet light transmission of DNA codes are geometrically structured water molecules – thus sacred geometry. Most important; it is via the medium of geometrically structured bio-water that the information encoded in your DNA is presently being transmitted. The medium of transmission for these new DNA instructions needs to be of a new order of sacred geometry. The old forms of pentagonal, and hexagonal clustered bio-water can be elevated as we step into our next evolutionary exploration. The DNA Activations can encode your cellular bio-water with new geometric forms, a higher octave of the Language of Light.

So after finding out this information, I felt my intuitive DNA Activations were being supported by scientific evidence and even more convinced at the power of this process.

After conducting numerous DNA Activations, I have had feedback of people's third eye being more open, their psychic abilities more alert, they are more aware of their life purpose, they have more x-ray vision, artistic talents fully blossomed instantaneously, their house fuses have blown or illnesses have disappeared. There have often been cases of feeling sensitive, over emotional, having flu-like symptoms and other temporary health issues. Oftentimes this is a healing crisis, that usually passes quickly and is a way of releasing old emotions and patterns. This shows the Activation has effected them clearing the way for more positivity and clarity. When I received the first Twelve Strand Activation, I responded profoundly and it all sounded so familiar, but I wasn't sure where from. When I tried to go to sleep after the session, my third eye opened dramatically and intensely for two hours as I watched fragmented memories of my past lives fly in front of me so rapidly, it was as if I was on an aeroplane. In the end, I asked my guides to stop it as it was all too much for me. Eventually it calmed down and I was able to sleep.

I was very fragile for a few days and some past emotions emerged I had forgotten about. The following week the lady who activated me tuned into me again and I was unknowingly re-activated to the point where I was seeing my organs and chakras light up with a lattice work of filaments of pulsating light. I had never seen this before and later found out that I witnessed my light body. As a bonus,

I received some X ray vision that I had no idea I had. I believe some of my latent talents returned with the activation. I later found out that when I activated someone and counted their DNA strands, I was receiving information from them, which allowed the release of inappropriate energy from their DNA, from either this life or a past life. My intuitive abilities had been upgraded and I was delighted!

Healing with the DNA

Stay relaxed as much as possible during the day and meditate for twenty minutes.

Go into meditation. Listen to your inner voice of your higher self. Imagine you are walking down some steps and count from one to ten and ask to be in theta level (brain wave patterns of 4 to 7 cycles per second.) and deep relaxation. Then, go through the body and tell each organ "I love you – you are healthy." Telling an organ that it is healthy every day, and believing it, you will create a more healthy body.

Go through the body from head to toe, including your etheric, physical, emotional, mental and spiritual bodies, plus the seven main chakras. Remembering your cells talk, ask the cells what is wrong emotionally, mentally or physically? You will receive an answer. You may be shown the problem or hear a voice that explains what is wrong. Trust your intuition. You can talk to the etheric bodies and chakras to sense if anything is wrong or blocked. Talk to the problem area if you have any and then say:

"I tune into my DNA Master Cell in my Pineal Gland to command a healing to be done on this problem (state what it is that you want done.) I activate All cells in my Body to receive this command. Thank you, thank you, thank you!"

Your guides may very well do all the work for you and you may be able to see how they work.

Sending light and colours into affected areas is also useful and is another form of cellular healing.

Surround and infuse the area with blue for cleansing, green for healing, pink to relieve pain, and white for universal strength. You can study what colours are good for each ailment and use them.

You may command "Relieve All Pain and Thank You!" This is useful for headaches, stomach aches, and the like.

You can also command all geometries that form that ailment to turn into centre, and create orbs of light that leave your body and grid.

Every organ, part of the body and illness has a tone. When performing the healing technique and illness is present, ask for the tone to come through that disempowers that virus.

As a illness becomes stronger, it may be necessary to use more than one tone, because a single tone may not work. Chromosomal diseases are caused by defective DNA cells that build walls around them, causing them not to communicate. When you destroy tumours and unhealthy cells, command the debris to be turned into healthy cells again or leave the body immediately.

Sound

I find that sound is an integral part of the DNA activations and I use many different chants during the process. I use different languages including Hebrew and Sanskrit, the first languages of the earth, plus other channelled words and chants. (Please see www.kryon.com for the channelled hebrew for the first twelve strands of the DNA.)

The concept of using sound to activate the DNA was known in ancient history and has been suppressed or lost due to our DNA being influenced or tampered with. This has greatly impeded human evolution. It is now time for all the original information to reappear and for us to attain the next stage in our evolution.

Certain frequencies and sequences of sounds have profound positive effects on the DNA. As these frequencies interact with the DNA, the encoded information is released, thus flooding the cells, which raises the vibration of the physical body, the energy body and beyond. The resonance engendered evokes a higher blueprint and a more expansive state of spontaneous regeneration. The more sensitive and aware you are to subtle energy fields, the more you will know of this truth.

In truth, we are energy beings, composed of various frequencies that intertwine to bring about the matter that forms the human body. Music, sound, modulation, harmonics and vibrations are fundamental elements emanating through and beyond the core-level subatomic of the third dimension. Sound and frequency are a very important part of feeding our body temple.

We all vibrate at a certain sound and note, yet, with the New Earth vibrations and our Ascension, we are changing our individual frequency. This could explain why there can be friction in current friendships, because as your "note" changes it causes dissonance with certain people, when there was previously more harmony.

Warm Up Laughter Exercise with Sound

Complete the following exercise with the sounds:
HUH
Stand with your feet apart and shoulders relaxed. You can move your body if you wish. Send a series of HUH, HUH sounds to your hips and pelvic area.
HO
Then send a series of HO, HO sounds into your belly area.
HA
Then send a series of HA, HA sounds to your chest and heart area.
HEH
Then send a series of HEH, HEH sounds to your throat.
HEE
Then send a series of HEE, HEE sounds to the top of your head.
Finally move from one sound to another at random and by this time you should be laughing hysterically!

Solfeggio Scale

The original sound frequencies of the Solfeggio scale were used in ancient Gregorian chants, and were "lost" centuries ago. The chants and the electro-magnetic frequencies of this scale was believed to enhance the spiritual experience of those hearing the notes.

The sacred Solfeggio Scale was rediscovered in 1974 by

Dr. Joseph Puleo through a series of events that some would call miracles. He was guided to the Pythagorean method of numerical reduction, which led to his deciphering the frequency of these notes as hidden in the text of the Biblical book, Numbers.

In order for DNA to be reprogrammed, the cell membrane must first be opened to allow the information access to the DNA. One of the ways to open the cell is by activating the receptors in the cell membrane so that the message the RNA picks up will be transmitted to the DNA. The frequencies of the Solfeggio notes can open the cell, allowing new programming to be imprinted on the DNA and is another way to activate your DNA.

The Six Solfeggio Frequencies include:
UT – 396 Hz – Liberating Guilt and Fear
RE – 417 Hz – Undoing Situations and Facilitating Change
MI – 528 Hz – Transformation and Miracles (DNA Repair)
FA – 639 Hz – Connecting/Relationships
SOL – 741 Hz – Awakening Intuition
LA – 852 Hz – Returning to Spiritual Order

You can buy tuning forks attuned to this scale if you so desire, to allow activation of yourself and others.

Sounds to begin your DNA Activation

Chants

OM – thirty three times (the first sound of the Universe)
YOD HEY VOD HEY – thirty three times (one of the sacred names of God)
I AM THAT I AM – thirty three times
KODOISH, KODOISH, KODOISH ADONAI 'TSEBAYOTH- thirty three times (Holy, Holy, Holy, is the Lord God of Hosts)
Then meditate for fifteen to twenty minutes to Gregorian Chants or use your tuning forks.

Chapter Six

AKASHIC RECORDS

Akasha

"Akasha" is the Sanskrit word meaning "aether" in both its elemental and mythological senses. In Hinduism Akasha means the basis and essence of all things. The source of everything that exists.

Akashic Records

The "Akashic Records" (Akasha is a Sanskrit word meaning "sky", "space" or "aether") is a term from theosophy denoting a collection of mystical knowledge encoded in a non-physical plane of existence. The Records are supposed to contain all knowledge, including all human experience, held within the Universe. The Akashic Records are metaphorically described as a library and are also likened to a universal computer or the 'Mind of God'.

Edgar Cayce was a very well known psychic channel in the twentieth century and was asked where he obtained the information in his readings, and he noted two main areas – the subconscious and what he called "The Hall of Records."

Cayce would put himself into a trance state for his readings, see himself as a dot of light outside of his body and travel through dark lonely spaces to reach the Hall of Records. An old man would then come to greet him and direct him to the information about the person for whom he was giving the reading. Cayce was shown a book, he said, because humans expect knowledge to come in this garb. He would be shown the aspect of the book, which would be most relevant at that time, to the person receiving the reading. More messages would be given at a later time once the person had applied the given information to their life.

Most people want their desires to manifest, but in fact, "getting what you want" is not necessarily the point of our existence. Cayce said we are co-creators with the divine. We've been granted free will and that free will allows us to choose how we will respond to people and situations.

As I mentioned previously, these responses and choices effect our life and if we have learnt our lessons, we begin to work with more positive choices easily and naturally.

Due to the quantum shift now available in the DNA, you can contact your Akashic Records. You can regain the ability to access the Hall of Records to attain knowledge and gifts you once had, allowing you to retrieve and use them again!

We can close the door to the Old Matrix thinking that these abilities are from past lives, and that they only existed in the past and not in the now. Yet, as we become more lifted into a quantum level of understanding, which will unfold naturally, we can access more and more of who we were and who we really are.

Akashic Field Theory

In the book *Science and the Akashic Field: An Integral Theory of Everything* by Ervin László, he posits that a "field of information" as the substance of the cosmos. Using the Sanskrit and Vedic term for "space," or Akasha, he calls this the "Akashic field" or "A-field." He suggests that the "quantum vacuum" is the fundamental energy and information-carrying field that informs not just the current universe, but all universes past and present and is termed the "Metaverse."

László describes how such an informational field can explain why our universe appears to be so fine tuned as to form both galaxies and conscious lifeforms, as well as why evolution is an informed, non-random, process. He believes that the hypothesis solves several problems that emerge from quantum physics, especially non-locality and quantum entanglement.

Akashic Record Clearing

We can say our personal Akashic Records are like the Encyclopedia Britannica, but instead of twenty six volumes, we could be talking about over three hundred thousand volumes depending on how many lives we have had! Also, we have to keep in mind we have probably been everything from the good to the bad, and everything in between.

The soul is linked to the Akashic records and emits energy with every thought, action or event that it engages in and leaves a permanent energetic blueprint. The Akashic Records are said to be located in the Hall of Records, located in the Akashic Plane of the Universe. The records are located beyond the subconscious and physical realms and found between the subconscious mind and the many mansions of God. One of the main reasons to access your personal Akashic Records is to view that which is holding you back and then to clear it.

Much like a computer, the records hold every programme of every emotion, thought, feeling and action from all events and from all lifetimes, whether we use the programme or not. In the clearing methods I utilise, I clear the main programmes from all lives and thus release you from that which is hindering your evolution. This occurs whether you are aware of the programme or not, although free will is still dominant and life lessons cannot be avoided.

There are two ways to access the Akashic Records. One is to use the help of an experienced channel who is able to view the records. The other way is to slowly train yourself to access the records through meditation and increased awareness. It is important to remember that one must begin this procedure with a prayer to ask permission from the Akashic Record Keepers and to give thanks to God, our soul guides, angels and masters.

When we want to access the Akashic Records, we cannot just view our life inordinately. We can only view what is given to us by asking specific questions to the keepers, or our guide, about our past and present. Then, we are given access to a "window" through which we see relevant scenes of our past life, parallel life or present life happening before us. Interestingly, we can also view future lives, which are dependent on our current thoughts and actions.

Akashic Record Clearing Meditation

Please choose a pattern, programme or habit you wish to release and clear and then say this prayer out loud.

"Dear Mother Father God, Akashic Record Keepers, Soul Record Keepers, My Soul Group, My Twin Flame, Guardians and Guides, please may I clear this chosen programme from my Akashic Records. Thank you, thank you, thank you. Amen!"

Close your eyes. Take three deep breaths.

Ask to see a scroll which symbolises the Akashic Records and Hall of Records. Try to see the scroll before you to access your personal Akashic Records. You may not see a scroll and you may be given a different symbol such as a hall or many mansions.

Say out loud:

"I call forth the Akashic Record Keepers to clear my chosen programme of "...................." Please clear now. Clear, Clear, Clear."

Wait a moment for this to be done. Take three deep breaths.

Then say out loud:

"Thank you Mother Father God, Akashic Record Keepers, Soul Record Keepers, My Soul Group, My Twin Flame, Guardians and Guides thank you for clearing my chosen programme. May it be so. Amen!"

Chapter Seven

SYMBOLS and DNA

"The soul never thinks without a mental image" - Aristotle

Lord Melchizedek – Channelled Symbols

One day during a healing session I was receiving, Lord Melchizedek came to me and related that there were One Hundred and Forty Four symbols to assist us with ascension and to activate our DNA. After the healing, I immediately sketched and channelled the first Twelve Symbols that are available in full colour upon request.

Ascended Archangels and Master Symbols

Some years later, I was inspired to produce another set of symbols and interestingly, with a pendulum. I tuned into each Archangel and Master and asked the pendulum to channel and help me draw the energy of each Being and to also choose the colours. The colours throughout the forty four drawings were purple and gold. I have included nine of these symbols and meditations, one for each Master.

The Twelve Melchizedek Ascension Symbols

Each symbol has a higher vibrational colour, mantra, sound, pitch and meaning. They are tools for the focus of energy, which allows your consciousness to expand, self-healing to occur and the ability to raise your vibration.

Meditation for each symbol:

Breathe in twelve times each:

Words at bottom of symbol and suggested colour

Symbol to 3rd Eye - expand it all around you

Sing the Mantra at top of symbol, on the note of the piano keyboard shown, or the note that you sing can also be intuitive

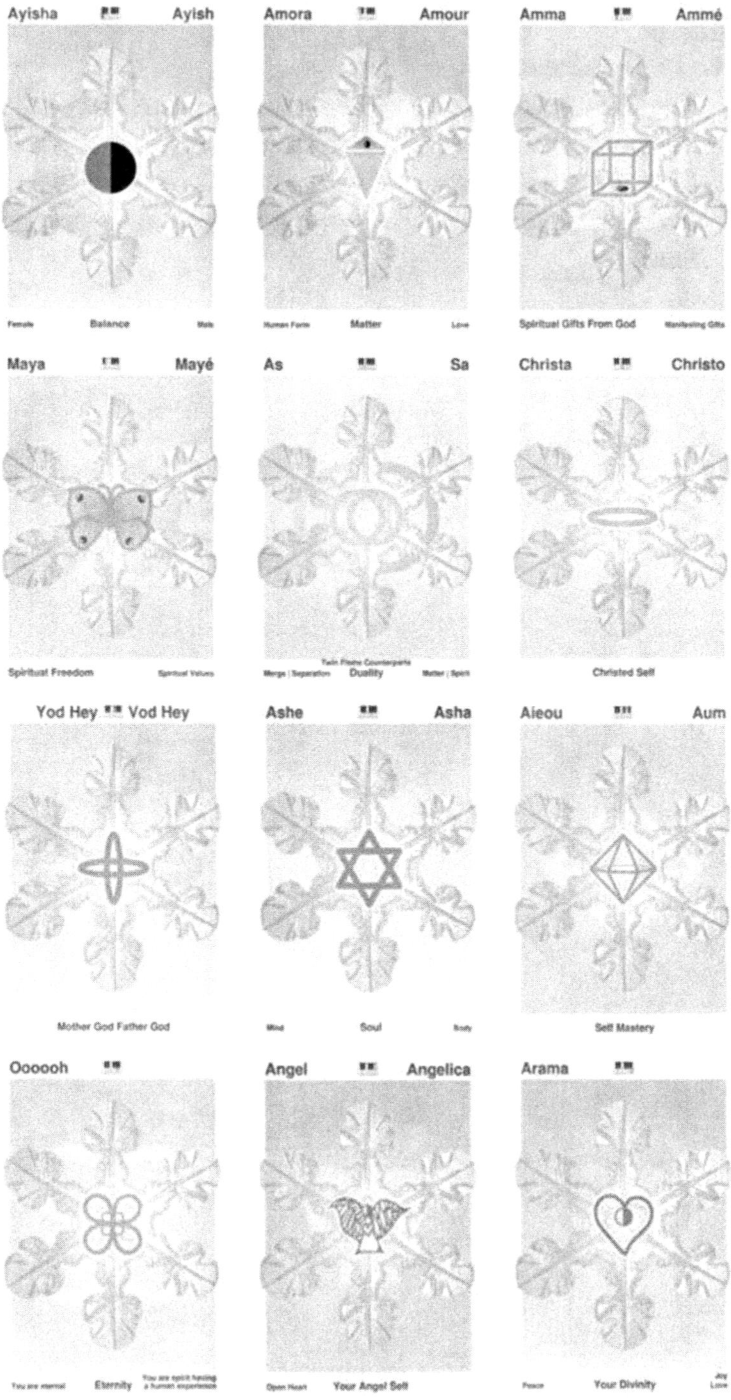
Ayisha
Ayish
Balance
Amora
Amour
Matter
Amma
Ammé
Spiritual Gifts From God
Maya
Mayé
Spiritual Freedom
As
Sa
Duality
Christa
Christo
Christed Self
Yod Hey
Vod Hey
Mother God Father God
Ashe
Asha
Soul
Aieou
Aum
Self Mastery
Oooooh
Eternity
Angel
Angelica
Your Angel Self
Arama
Your Divinity

Colour Pink

Meditation

Breathe in twelve times each:

Words at the bottom of symbol and suggested colour

Symbol to 3rd Eye - expand it all around you

Sing the Mantra at top of symbol, on the note of the piano keyboard shown, or the note that you sing can also be intuitive

Ayisha **Ayish**

Female **Balance** Male

Colour Pearlised Pink
Meditation

Breathe in twelve times each:

Words at bottom of symbol and suggested colour
Symbol to 3rd Eye - expand it all around you
Sing the Mantra at top of symbol, on the note of the piano keyboard shown, or the note that you sing can also be intuitive

Amora **Amour**

Human Form **Matter** **Love**

Colour Peach
Meditation

Breathe in twelve times each:

Words at bottom of symbol and suggested colour
Symbol to 3rd Eye - expand it all around you
Sing the Mantra at top of symbol, on the note of the piano keyboard shown, or the note that you sing can also be intuitive

Amma **Ammé**

Spiritual Gifts From God **Manifesting Gifts**

Colour Pearlised Peach
Meditation

Breathe in twelve times each:

> *Words at bottom of symbol and suggested colour*
> *Symbol to 3rd Eye - expand it all around you*
> *Sing the Mantra at top of symbol, on the note of the piano keyboard shown, or the note that you sing can also be intuitive*

Maya **Mayé**

Spiritual Freedom **Spiritual Values**

Colour Lemon
Meditation

Breathe in twelve times each:

Words at bottom of symbol and colour

Symbol to 3rd Eye - expand it all around you

Sing the Mantra at top of symbol, on the note of the piano keyboard shown, or the note that you sing can also be intuitive

As **Sa**

Twin Flame Counterparts

Merge | Separation **Duality** **Matter | Spirit**

Colour Pearlised Lemon Meditation

Breathe in twelve times each:

Words at bottom of symbol and colour

Symbol to 3rd Eye - expand it all around you

Sing the Mantra at top of symbol, on the note of the piano keyboard shown, or the note that you sing can also be intuitive

Christa Christo

Christed Self

Colour Pale Green
Meditation
Breathe in twelve times each:

Words at bottom of symbol and colour
Symbol to 3rd Eye - expand it all around you
Sing the Mantra at top of symbol, on the note of the piano keyboard shown, or the note that you sing can also be intuitive

Yod Hey Vod Hey

Mother God Father God

Colour Pearlised Green
Meditation

Breathe in twelve times each:

Words at bottom of symbol and colour
Symbol to 3rd Eye - expand it all around you
Sing the Mantra at top of symbol, on the note of the piano keyboard shown, or the note that you sing can also be intuitive

Ashe **Asha**

Mind **Soul** **Body**

Colour Turquoise
Meditation

Breathe in twelve times each:

Words at bottom of symbol and colour
Symbol to 3rd Eye - expand it all around you
Sing the Mantra at top of symbol, on the note of the piano keyboard shown, or the note that you sing can also be intuitive

Aieou **Aum**

Self Mastery

Colour Pearlised Turquoise
Meditation

Breathe in twelve times each:

Words at bottom of symbol and colour
Symbol to 3rd Eye - expand it all around you
Sing the Mantra at top of symbol, on the note of the piano keyboard shown, or the note that you sing can also be intuitive

Oooooh

You are eternal **Eternity** You are spirit having a human experience

Colour Lilac
Meditation

Breathe in twelve times each:

Words at bottom of symbol and colour
Symbol to 3rd Eye - expand it all around you
Sing the Mantra at top of symbol, on the note of the piano keyboard shown, or the note that you sing can also be intuitive

Angel **Angelica**

Open Heart **Your Angel Self**

Colour Pearlised Lilac
Meditation

Breathe in twelve times each:

Words at bottom of symbol and colour
Symbol to 3rd Eye - expand it all around you
Sing the Mantra at top of symbol, on the note of the piano keyboard shown, or the note that you sing can also be intuitive

Arama

Peace   **Your Divinity**   Joy
Love

Ascended Archangels and Master Symbols

Here are nine examples of the forty four symbols channelled using the pendulum. Each of three symbols per page have a meditation beside them. You can concentrate on each symbol for about twenty minutes and call in that particular master with the meditation.

Lord Azrael

Keeper of Twin Flames
Carrier to the Stars
Caretaker of Angels
I help you reunite with your Twin Flame. Being One with your Twin Flame in Soul is the ultimate position of Divine Love. To know your higher aspects together as one, you will find your wholeness and the true beauty of your divinity.

Meditation

I ask permission from God to meet my Twin Flame in Soul to make me whole again

Saint Francis

Serves all Planetary Life Forms

Meditation

Lord, make me an instrument of Thy peace;
where there is hatred, let me sow love;
where there is injury, pardon, where there is doubt, faith;
where there is despair, hope;
where there is darkness, light;
and where there is sadness, joy...

Lord Akhenaton

Assists Self Mastery
I help thee find your true path through the wisdom of ancient texts and the writings of the masters. I also inspire thee to become more aware of your master within and guide your ascension.

Meditation

I call Lord Akhenaton to guide me to wisdom through the written word and to be in touch with my mastered self within.

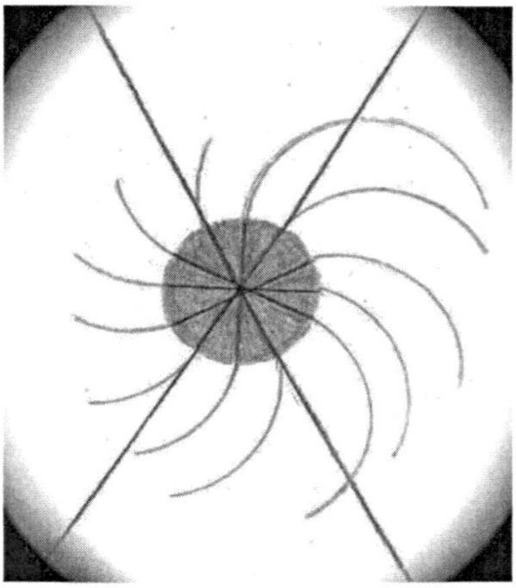

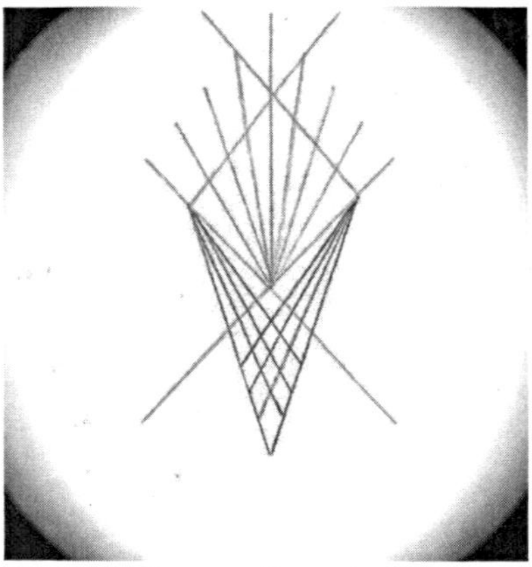

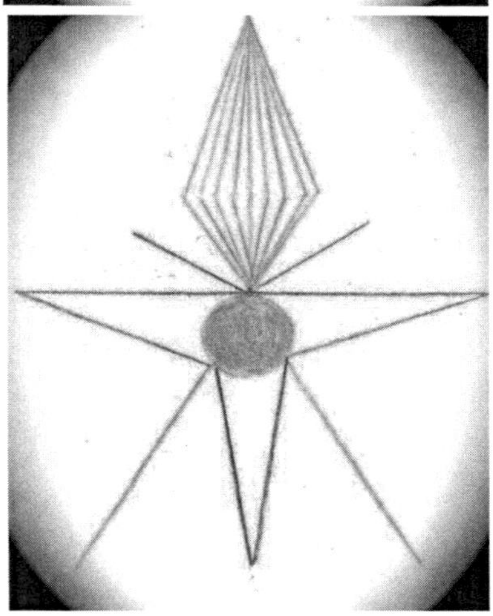

Lady Qwan Yin
Mother of Compassion
Magnified Healing
She fills you with five virtues:
Mercy, Modesty, Courage, Justice and Wisdom
Meditation
I fill my whole body with the Essence of Qwan Yin

The Essenes
Brotherhood and Sisterhood
The Essenes in the higher offer many teachings of ancient universal wisdom, as a sacred service to humanity.
Meditation
I ask the Heavenly Father and Earthly Mother that my soul becomes as perfect as Heavenly Father's holy spirit is perfect, and my body becomes as perfect as the body of the Earthly Mother is perfect.

Lord Sananda
The Risen Christ
Unconditional and Divine Love
Twin Flame Unity
I come this day and every day and light a candle for thee, to show thee the way Home. When you follow this light you will remember your divinity.
Meditation
I ask Lord Sananda, The Risen Christ to watch over me, my family and friends, and all those who live upon the earth. I ask thee to answer my heartfelt questions and in the silence I will receive divine wisdom. Amen

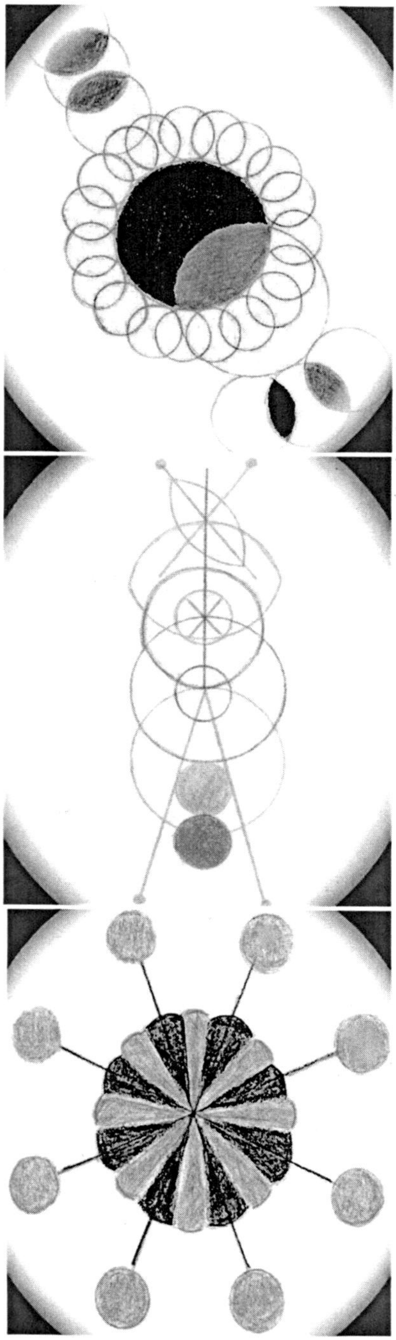

Lord Buddha
Enlightenment
Perfect methods for overcoming suffering and attaining happiness
Meditation
May everyone be happy, May everyone be free from misery,
May no-one ever be separated from their happiness,
May everyone have equanimity, free from hatred and attachment.

Lord Serapis Bey
Inspires Personal Initiations and Ascension
Balances and Activates Artistic Beauty
I have come to walk by thy side, to show you the way forward into more Ascended Light, to share that walk with you until you find your soul knowing that we are all One.
Meditation
I ask Lord Serapis Bey for guidance upon my path towards my Ascension to the Light and I know he walks softly beside me.

Lao Tsu
Teacher of Harmony and Philosophy
I have just three things to teach: simplicity, patience, compassion. These three are your greatest treasures.
Meditation
I ask Lord Lao Tsu to show me how to live in the moment and master my self, knowing that all answers lie within me.

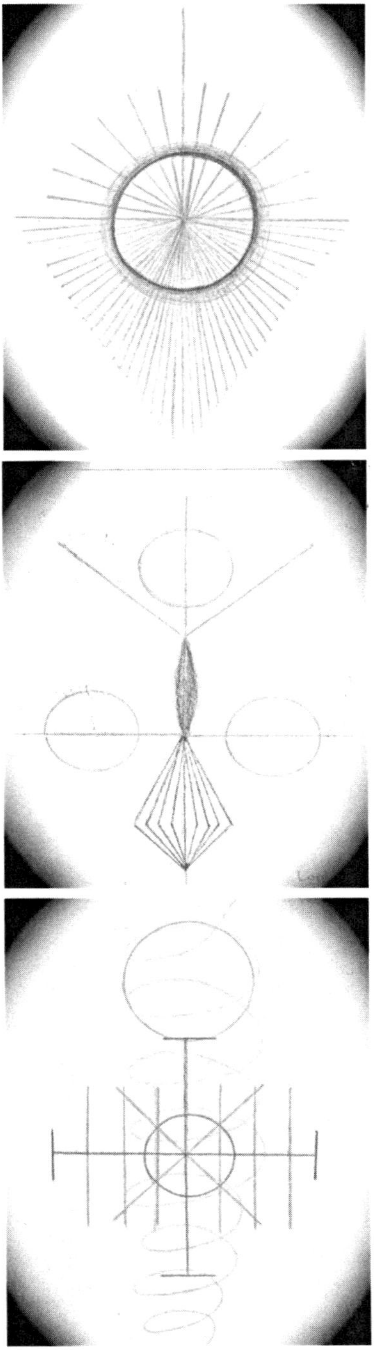

Chapter Eight

ORBS AND TOROIDS

DNA is crystalline in nature and contains All That Is. In the DNA Activation process I include sacred geometry, the Flower of Life and the Merkaba. Sacred geometry is the underlying form of life within this Universe. The Flower of Life, which is nineteen circles intersected, holds all sacred geometric shapes and it is said that our DNA and RNA are born from this holographic pattern. The finest particle of our atomic cell structure has this pattern within it.

The Merkaba is a rotational vehicle of light, body and spirit used to obtain higher consciousness and God. Please read Drunvalo Melchizedek – *The Ancient Secret of the Flower of Life* Volume I and II, for further in-depth information.

In 2009 I was given a powerful message that I was to activate the "Rainbow Light Body" a new energetic level of our bodies. I was not sure what that meant and there was very little information I could find on the rainbow light body. Anyhow, during a healing session a few months later, New Earth Orbs some half the size of my body, in glorious new colours and Divine Intelligence, appeared around me and merged with me. I was immersed in indescribable colours, so I presumed this was the Rainbow Light Body they were delivering. It was blissful as they are also of Divine Love and they visited me afterwards on a regular basis for a few weeks. At one point some orbs visited me very playfully and joyfully like children, and gave me a sense they had something to do with the New Earth children to be born. They reminded me that all is Love and being Love is All that matters! I became very inspired and intrigued by this.

The Circle becomes the first enclosed form of Sacred Geometry. It is unity and oneness, a two dimensional manifestation of one pointedness and undivided God Mind. The Circle is the All and the essence of the mandala which holds All.

The New Earth Orbs represent new creations of a higher consciousness now in the process of preparing the New Earth vibrations. This is part of the Cosmic and Earth Ascension, and our Ascension in that we are able to keep the body. This process has never been achieved in this manner on earth.

Orbs =

Orbs can be any of the following:

- A sphere or spherical object
- A celestial body, such as the sun or moon
- Archaic. The earth.
- One of a series of concentric transparent spheres thought by ancient and medieval and astronomers to revolve about the earth and carry the celestial bodies
- A globe surmounted by a cross used as a symbol of monarchial power and justice
- An eye or eyeball
- Archaic. Something of circular form; a circle or an orbit
- Archaic. A range of endeavor or activity; a province

Orbs have often appeared in photographs or around haunted houses, but these are not the Orbs I wish to speak of. I wish to speak of the New Earth Orbs.

Orbs can appear in a rainbow of colours. Barely physical, their bodies can be described as bubbles of energy. Scientists, who have researched Orbs, believe their body is made of dense groupings of charged particles called plasma. This could mean that Orbs are the simplest sentient beings in the Universe.

Recently at an ashram in India the phenomenon of Energy Orbs has become more and more visible. These beautiful representations of planetary energy are becoming manifest in response to the ever-growing level of consciousness present at this time. Feeling their presence and working with them during training and meditating, allows you to connect with your highest potential, free from the limits of the mind. When you are connected, through feeling, to these energy orbs, spinning occurs throughout the body, and while held in the safety of ones own orbit, it can feel like dancing through the universe.

We are now shifting our vibration into the New Earth Orbs, which help rebirth the sacred geometric patterns held within the cellular. This, in turn, helps in releasing dense enrgy from our Old Matrix Cells body. The Orbs give us more flow into Divine Joy within the body and the ability to access latent higher-dimensional gifts and talents.

We are now re-activating our two chromosomes of the Rainbow Light Body and the New Earth vibration. These two chromosomes represent the New Earth Mother/Father God, remarried at the Magnetic levels of Pure Heart Love. We can thus manifest, create and change instantly and at will. DNA can be called Quantum DNA, which is the energy intelligence that connects to multi-dimensions. We have DNA that gives us multiple potentials and we have billions of strands still to be acknowledged and activated.

It is said we will be able to create through new higher-vibrational colours, sounds, symbols and thoughts. Once we are activated in the Rainbow Light Body through these New Earth Orbs, we have a Peace and Creativity unfelt and unknown before during any earthly incarnations. This is also unknown to our cellular records. This ascension is happening for the first time in this manner and the body does not have to die in order to ascend, as has happened during previous lifetimes. In other words, our physical body is ascending with our higher consciousness and we do not have to leave it behind. Even our guides do not know how it is going to effect us on a daily basis. This is why we may have more of the ups and downs we have to deal with emotionally and physically, as we change our vibration. We are now welcoming the New Earth.

Introduction to the New Earth Orb Activation

Close your eyes and relax.

Call in your Twin Flame at a soul level and Lord Metatron, who is in charge of the New Earth changes and his Emerald Flame.

Surround yourself in His Emerald Flame.

Now invite the New Earth Orbs to be revealed to you in their magnificent colours and their new creations.

Breathe in all the colours of the rainbows and ask for all the new colours to reveal themselves their opalescent and pearlised qualities.

Allow your body, cells and DNA to absorb the new orbs and colours.

When you are ready come back to awareness of your body, keep your feet grounded and open your eyes.

Toroid =

- the doughnut-shaped object enclosed by a torus
- a ring-shaped surface generated by rotating a circle around an axis

It is said, in the beginning of our Universe, matter formed into toroidal formations some 311.040 trillion years ago.

The shape of the Universe looks like a toroid. Within any 4D toroidal shape, there are mathematical relationships that are unique. Cutting edge scientists are discovering that toroids may be the core of form, from the infinitely large, to the infinitely small, including our DNA itself.

While working with the new energies more recently, I have had moments when my Heart Chakra turned into a toroid of colour and energy, and then in turn, my entire body itself turned into a toroid. I felt I could have disappeared during this experience and this may be a tool that enables us to teleport!

Toroid Meditation

Close your eyes and relax.

Place a multi-coloured mandala on your heart. See all the beautiful patterns and vibrant colours. Then imagine a doughnut that is energetic and moves into the centre and around and around itself. This is how the Universe and DNA moves. Imagine this doughnut and mandala moving like a kaleidoscope into the centre of your heart, then out the back of your heart in spiraling doughnut shapes.

These doughnuts start to form on your wrists, knees and ankles, which move in the same colourful way. This movement begins to become one magnificent emanation of movement, patterns and colour. The spinning toroid becomes the essence of creation and Divine Love. You also become this movement and you become your Christed Self along with your body. You become Divine Love and Divine Creation and are One with your Twin Flame and God. You are your Rainbow Light Body and New Earth Body.

Before you open your eyes, remember to grow roots over your feet to fully ground yourself back in your body. Amen.

Chapter Nine

GOD'S EYE

It is said that before the fall of Atlantis, we breathed differently and utilised "prana" running through our pranic tube. In Hindu tradition, prana is the life sustaining force which pervades all living organisms and the universe. The pranic tube is crystalline in form and runs vertically, mid-body, from just below our feet to just above our head. It is about the diameter of the circle you make when you hold your thumb to your third finger. The prana is said to have worked with the Pineal Gland, located almost in the centre of the skull. Its original size was the size of a ping-pong ball but has now shrunk to a shrivelled pea. We have forgotten how to breathe prana into the pineal gland and thus it has lost its full use.

It is also thought that the Pineal Gland is a real eye used to look up to God and holds sacred geometries along with the understanding of exactly how the universe works. All these mem-ories are suppressed as we now breathe through our nose and mouth as the prana bypasses the Pineal Gland. Thus, we have fallen in consciousness for many thousand of years.

I believe that the new energies of the New Earth that are now available, will re-open our God's Eye, or the Pineal Gland eye. I know this as I have personally and completely experienced this opening, totally out of the blue, during the summer of 2009. One night, Lord Akhenaton, in soul, visited me, and was in charge of opening my Pineal Gland as a real eye. I then remembered a time in Egypt when I was very tall and when all these capabilities were available to us. I was able to see upwards, sideways over the ears, and back and front of the third eye. There was a link of energy to the earth from the Pineal Gland. I felt I had had brain surgery and my brain literally ached for days afterwards!

Lord Akhenaton Pineal Gland Activation
As channelled by Lord Akhenaton

Beloved Ones,

In these days of resurrection of your true capacity as a human being and in remembrance of your true divinity as a human being, a soul with a physcalised component, we honour you this day with a technique to ignite your true third eye or God's Eye.

Extreme care must be taken not to overload the body and brain area all at once so this a step by step Pineal gland initiation.

Please call your Twin Flame and Lord Akhenaton and his team to oversee this activation.

Step One
Close your eyes and receive a spiralling of universal energy from the heavens through your crown to below your feet and then a spiralling upwards from below your feet to above your head. Repeat this over and over.

Step Two
Ignite all chakras with a deep breath of light through your pranic tube which is approximately two inches wide, running from your crown down your spine to below your feet. Ask the light to pass through each chakra: base, abdomen, solar plexus, heart, throat, third eye, and crown.

Step Three
Open your Pineal Gland as an eye in the centre of your skull

- *looking up*
- *looking down*
- *looking out your forehead*
- *looking out the back of your head*

Step Four

Repeat the gaze projecting light in these four directions in the form of beams one quarter of an inch wide.

Step Five

Divide these four beams with a light beam one quarter of an inch wide coming out your right ear and left ear from the pineal gland.

Step Six

Meditate opening these beams and continue the spiralling of energy up and down your body for thirty minutes. Give yourself to God.

Ask your Twin Flame to ignite two orbs over your Pineal Gland, then to intersect them allowing the formation of New Creation. This is also called a Vesica Piscis position and Twin Flame position.

Your New Earth body is to be ignited in its perfected form through the centre of the Vesica Piscis, like an explosion or re-birth, and through your God's Eye, which is your 3rd eye in the Pineal Gland in a fully awakened state.

Slowly come back into your body and ground yourself. Amen

The Thymus Gland

The Thymus Gland is a pinkish-gray organ that is located in the upper chest under the breast bone. It is the New Mother/Father God-Heart and can be seen as heart shaped. It is a doorway to the Hall of Records, Teleportation, Manifestion, Telepathy, X-Ray Vision, Precipitation and all latent talents and skills.

The thymus gland is growing in size just like the pineal gland. It is a direct doorway to heavenly realms and New Earth vibrations. See it as a Heart of Gold and as an energetic axis point to all dimensions. It is pivotal to control movement of cells to the New Earth vibratory rate.

Thymus Gland Activation

First start by tapping lightly your chest area above the heart chakra in circular movements. First one way and then the next.

Take three deep breaths into your thymus.

Fill it with lavender.

Take another three deep breaths into your thymus.

Fill it with rainbow colours.

Take another three deep breaths into your thymus.

Fill with with gold.

See the thymus expand healthily into its original form.

Breathe in the New Earth Mother/Father God vibration into your thymus.

Repeat these statements out loud:

"I activate all cells in my thymus to be activated to the New Earth Body vibration and open the door way to the Halls of Records, Teleportation, Manifestion, Telepathy, X-Ray Vision, Precipitation and all latent talents and skills. I activate all my gifts now. May it be so."

Chapter Ten

EVERYTHING COMES BACK TO LOVE

The New Paradigm

A channelled message from a spokesperson of the family of God and Earth

Dear Beloveds,

You will see a new paradigm in your life time. The veils will be lifted and you will walk the earth amongst those of love and compassion. This love will touch all, especially in the western world. The lighted ones will take charge of the minds of those in power as new people of integrity step forward. The lighted ones, being the master within, in agreement with Gods force of love and light. Beloveds, this change is the greatest contrast from dark to light you have ever known on earth. Not only will you see justice, but you will see enlightenment. For the forces of light are now amongst and amidst you, standing right in front of you through the children, teenagers and early twenty year olds.

The time is upon you and your prayers are answered - indeed all ancestors of the at-onement with Mother Earth and Father God will return en force and you will live in tune once more with that which is most important - being the caretakers of Mother Earth and the keepers of one another.

Simplicity will replace greed, humility will replace power, and joy will replace oppression. You will see this come to pass and because of your presence, dear Masters of the Light, you have set the ground work and the foundation to make this occur

Rejoice, rejoice, as the final days of the old will be replaced with all that is of nature, spirit and soul. You will see great changes and you will sigh a deep sigh of relief. You are the children of the future and with that comes optimism, faith and rebirth.

God bless you all!

To help you on a daily basis and to tune into the higher realms you can use this channelled prayer:

Divine Love Daily Prayer

I AM A DIVINE CHILD OF GOD
I deserve Divine Love and Light from Mother/Father God
I deserve Divine Love and Light from my Twin Flame
I deserve Divine Love and Light from my Soul Group
I deserve Divine Love and Light from the Masters, Archangels, Angels and Guides
I deserve all my Higher Aspects to return to me
I deserve all aspects of my Twin Flame to return to me
I am a Great Soul having a human experience
I deserve everything, success in all areas of my life is my divine heritage
I am Whole
I am Divine Love
I am my Godself
I am my Christed Self
I am my Angel Self
I AM ETERNAL

I command to set myself free of all third dimensional burdens, judgements, contracts, karmic and negative patterns and I command all the negativity I have been holding to never to return.
AND SO IT IS!

Meditation to Be with Your Christed Self

Close your eyes and relax.

Connect to Mother Earth through your roots extending

downward from your feet and bring up her healing energies through your chakras. One by one, spin them front and back of the body – the base, abdomen, solar plexus, heart, throat, third eye, crown, along with the chakras above the crown a few inches above each other. Open your soul-flame at the same time through the chakras – Clarity, Harmony, Balance, Eternal Peace, Divine Purpose, Joy, Transformation, Divine Love, and your Christed Self.

Repeat out loud to open your top chakras above your crown:
I activate my soul flame above my crown
I Am Clarity
I Am Harmony and Balance
I Am Eternal Peace
I Am Divine Purpose and Joy
I Am Transforming
I Am Divine Love
I Am my Christed Self
And So It Is

You then find yourself in a beautiful realm and on pristine land where your soul group live. You now call Lord Sananda. The many singing angels around Him step through first and then you see this very bright light as He reveals himself. He has a candle that is lit especially for you to show you the way forward.

His presence activates you to be in your Christed-Self and Christed-Body. You are standing much taller than usual. You are standing as a being of Light. You are observing yourself as a being of Divine Love and you are one with your Christed-Self.

Stay in this position for fifteen minutes. See if any messages are given to you, or if anyone else visits you.

Before you open your eyes, you can ask for all chakras to close like flowers and that your feet are grounded by the roots growing over them, holding you to Mother Earth.

Divine Love
Channelled by Dr Kruger

Many of you think that life consists of moment by moment by moment, and you see things in a linear fashion upon the earth, yet, from our stance there is no moment by moment, there is only being and doing. We are very busy here, doing and being. We are very busy helping other souls. We are very busy helping the angelic realms. We are servants of God and we are busy on many, many levels. We are busy in all our aspects emanating from our soul. We may visit you in one aspect one day and another aspect another, depending on your mood and that which you need. We are busy learning about our own soul and that which we can accomplish whilst we are on this side of the veil.

For some of us it is not important to reincarnate quickly and we remain in soul awareness for long periods of time, often for hundreds of your years. We grow by understanding our Divinity and by understanding God. It is not easy to fathom, for it is an eternal learning. And just as you are changing vibrations, so do we, therefore it is never ending. How can we ever stop learning? The most important aspect of our learning here is to become Divine Love.

Divine Love sits at the Heart of your Soul. You were born from this Divine Love and in your original birth and the first day of being given intelligence and a mind, you were of born of God. Some of you consciously remained born of God and some of you went off course. Yet, the test whilst in a human body is always to remember God, to remember your Self, to remember your Heart that lives within your Soul, and to recognise Divine Love.

In Love and Appreciation of who You are,
Dr Kruger

Meditation to reconnect to the Heart of the Soul

Close your eyes and relax.

Connect to Mother Earth through your roots extending downward from your feet and bring up her healing energies through your chakras. One by one, spin them front and back of the body – the base, abdomen, solar plexus, heart, throat, third eye, crown, along with the chakras above the crown a few inches above each other. Open your soul-flame at the same time through the chakras – Clarity, Harmony, Balance, Eternal Peace, Divine Purpose, Joy, Transformation, Divine Love, and your Christed Self.

Repeat out loud to open your top chakras above your crown:
I activate my soul flame above my crown
I Am Clarity
I Am Harmony and Balance
I Am Eternal Peace
I Am Divine Purpose and Joy
I Am Transforming
I Am Divine Love
I Am my Christed Self
I Am my God Self
And So It Is!

You now pour peace into your auric field - feel it spread out larger and larger. You get a sense of your greater-self, larger and larger, taller and taller. There, at the centre of your greater-self, lies the Heart of the Soul.

The Heart of the Soul is golden and heart shaped. It is large enough to encompass your human body. See yourself sitting comfortably in your own Heart of the Soul. You are completely surrounded by Golden Light and Divine Love.

Remain there for fifteen minutes or so in Peace and in recognition of all the qualities of your Soul's Heart.

Before you open your eyes you can ask for all chakras to close like flowers and that your feet are grounded by the roots growing over your feet holding you to Mother Earth.

Amen and Blessings to You.

Afterword

I believe that we are being made ready for the greatest shift we have ever witnessed in this lifetime and perhaps in all lifetimes.

Please prepare yourselves for this amazing time ahead with self honesty, and be in observation and detachment mode to that which we are witnessing in the matrix and upon the earth. I hope that you have gained some thoughts and ideas about how to move to your next level of spiritual awareness through my writings, and even though we can feel very alone upon our chosen path, remember we are all One at Soul.

I am wishing you joy, laughter, wisdom and tons and tons of love.

Blessings to you,
Jayne Chilkes

Channelling from Lord Azrael/Azraella

Dear Ones,

If only you knew how much you are loved and when you fall into darkness or those around you are fallen, you are not lost to us. You are still loved, still honoured, still adored by all your soul group and keepers of the Light. When you call us we will come. We will not desert you. You are protected if you allow us to protect you, you are helped if you allow us to help you and you can feel our Love if you so allow.

God bless you dear children of the Light for you are ALL beautiful beings of the heart of God.

You are particularly dealing with extremes at this time of great change, the good will seem greater and the dark will seem darker. Yet two sides of the same coin, we understand that there is always a choice and those who do not find themselves in the New Earth energies will live together as if on a seperate realm of earth.

There will be two earths yet not that which a human mind can totally understand. There will be two realms of earth and those who have chosen will walk the New Earth with all that will create a healthy life for all and Mother Earth herself.

The separation will become so obvious that you will not tolerate the old earth ways nor be able to live amongst them. It will be arranged that you are surrounded with that which resonates to maintain the new. Miracles occur for you as you find yourself helping to create mass joy and peace on earth. You are placed perfectly in your life plan not just for you but for Mother Earth and humankinds Ascension.

Appendix I

Responsibility to Self

Channelled by Dr Kruger

Beloveds,

It is of great importance that we speak to you this day about self responsibility. You may have been working very hard on yourself and helping others and also, you may have done all you thought you could do in therapies and healings. You may have released programmes and issues over and over but still you say to us "I am not done yet, there is still more work to do!"

In the reality of your third dimension and in a physical body you have had so many incarnations and lives of drama, conflict and challenges. There are some patterns left in the grooves of your memory banks that seem to recur over and over and do not want to be released or even be acknowledged by self.

This time round is your greatest opportunity to conquer the wheel of repetitive programmes of your own nature that now for the first time you know for sure, no longer serve you. Yet as much as you try you say to us, "they never seem to disappear."

You have to look deeper within to say to self, "I admit I enjoy the drama of the past, and my body enjoys the feeling of drama of my old behaviours. I enjoy making a mountain out of a mole-hole because then at least one other person might hear me and might listen to my pain!"

Peace seems boring to you, God seems distant from you, and you feel confused, yet at least you felt something and at least probably your partner, parent or close friend was forced to listen to you.

So the option is dear ones, let the Light take your pain, let

God listen to your issues and let a reservoir of healing rays pour through you each day. Let your cells rejoice in love and let go of the past as you rise in vibration and enjoy rising, instead of tripping over self over and over again.

We can assure you those vibrations now invisibly available to you are helping you resolve the deepest of grooves of patternings that you thought would never disappear. Nothing will be so difficult again as you walk through the doorway of the New Earth energies. This will become more and more obvious as the months pass by.

You will smile for no reason, you will laugh more and take life lightly, as the worries of the past dissolve and the switch is finally switched "on." These words are not said lightly. This is a Truth. You will walk in a physical body of Light.

May there be Peace in your Heart along your journey, may you take full responsibility for your actions and reactions and may you find your doorway of healing, to explore the truth of all that holds you back to your own inner Light. When you ask us for help we will come to show you how to move through your issues to the other side and practice Joy instead of hardship, Love instead of emptiness, and find Peace more enjoyable than dramas.

Many blessings
Dr Kruger and Healing Team

Breinigsville, PA USA
10 November 2010
249080BV00003B/283/P

9 781935 105503